Transcriptomics from Aquatic Organisms to Humans

Transcriptomics from Aquatic Organisms to Humans

Edited by
Libia Zulema Rodriguez-Anaya
Cesar Marcial Escobedo-Bonilla

CRC Press
Taylor & Francis Group
Boca Raton London New York

CRC Press is an imprint of the
Taylor & Francis Group, an **informa** business

First edition published 2022
by CRC Press
6000 Broken Sound Parkway NW, Suite 300, Boca Raton, FL 33487-2742

and by CRC Press
2 Park Square, Milton Park, Abingdon, Oxon, OX14 4RN

CRC Press is an imprint of Taylor & Francis Group, LLC

Library of Congress Cataloging-in-Publication Data
Names: Rodríguez-Anaya, Libia Zulema, editor. | Bonilla, Cesar Marcial Escobedo, editor.
Title: Transcriptomics from aquatic organisms to humans / edited by Libia Zulema Rodríguez-Anaya, Cesar Marcial Escobedo Bonilla.
Description: First edition. | Boca Raton : CRC Press, 2022. | Includes bibliographical references and index.
Identifiers: LCCN 2021027685 (print) | LCCN 2021027686 (ebook) | ISBN 9781032065168 (hardback) | ISBN 9781032079943 (paperback) | ISBN 9781003212416 (ebook)
Subjects: LCSH: Medical genetics--Research--Methodology. | Aquatic animals--Diseases--Genetic aspects. | RNA--Analysis. | Genetic transcription.
Classification: LCC RB155 .T724 2022 (print) | LCC RB155 (ebook) | DDC 616/.042072--dc23
LC record available at https://lccn.loc.gov/2021027685
LC ebook record available at https://lccn.loc.gov/2021027686

ISBN: 978-1-032-06516-8 (hbk)
ISBN: 978-1-032-07994-3 (pbk)
ISBN: 978-1-003-21241-6 (ebk)

DOI: 10.1201/9781003212416

Typeset in Palatino
by SPi Technologies India Pvt Ltd (Straive)

Contents

Contents

Preface

The field of molecular biology has experienced impressive technological advances in the last two decades, which has allowed it to unravel an array of novel mechanisms in many fields of the biological sciences. Transcriptomics is one of the novel molecular and sequencing technologies collectively known as "OMICS", including fields such as genomics, metagenomics, proteomics and metabolomics, which developed with the arrival of next-generation sequencing (NGS) technologies.

Transcriptomics allows analysis of the complete RNA content in a cell in order to elucidate the processes that activate and/or shut down these molecules. Transcriptomics makes it easier to analyze and understand massive amounts of information obtained from the complex relationships between external factors and molecules involved in any particular interaction.

As this technology becomes widely available in different research areas, better understanding of molecular functions and interactions are uncovered. This is the case for studies dealing with external and internal factors influencing health, as well as those involved in different aspects of host-pathogen interactions. These type of studies have been done using various models, including human, fish and invertebrates.

This book represents an up-to-date account of the applications of transcriptomics in different aspects including physiology, health and *de novo* genome sequencing. Such studies have been done in various organisms ranging from human to aquatic organisms of commercial, nutritional and/or ecological importance, including wild and farmed species of fish and marine shrimp.

The book is arranged into two parts: the first deals with the application of transcriptomics in aquatic organisms such as shrimp and fish. Chapter 1 discusses the application of different transcriptomics tools throughout time to unravel differential gene expressions in different organs of farmed shrimp upon different environmental conditions. Chapter 2 focuses on the application of transcriptomics in host–pathogen interaction in crustaceans. Chapter 3 presents the use of transcriptomics tools to understand different aspects of the physiology of fishes that may be useful in domestication, and selective breeding for potential aquaculture use. It also discusses the use of transcriptomics in ecological studies involving subjects such as adaptation and evolution.

The second part deals with aspects of human health, including noncommunicable human diseases and human infectious diseases. Chapter 4 presents the genetic and external factors contributing the presence of noncommunicable diseases in human populations such as cancer, diabetes, respiratory and cardiovascular diseases. The application of transcriptomics is widely used for the identification of biomarkers and transcripts related to the development and possible treatment of each disease. Chapter 5 examines the use of transcriptomic studies and RNA-Seq in the role of miRNAs on different types of diseases in humans, such as cancer, cardiovascular and neurodegenerative diseases, and metabolic syndromes. It also explores the role of miRNA as biomarkers of disease and as therapeutic agents, and its role in RNA editions. Chapter 6 discusses the mechanisms of pathogen–human interactions in infectious diseases caused by bacteria, fungi and viruses, including the coronavirus causing the current pandemic. The studies using transcriptomics focused on infectious diseases consider the molecular relationship between the host and infectious agent through transcriptional profiling and differential expression of genes during infection.

The contributors of this book have experience in their respective research areas. Their expertise includes next-generation sequencing techniques and transcriptomics, applied on subjects such as shrimp aquaculture and pathology, fish aquaculture and ecology, human non-communicable diseases and human infectious diseases.

Contributors

José Luis Acosta Rodríguez
Departamento de Biotecnología
 Agrícola
Instituto Politécnico Nacional–
 CIIDIR Sinaloa
Guasave, Mexico

Juan Diego Cortés-Garcia
Facultad de Medicina
Universidad Autónoma de San Luis
 Potosí
San Luis Potosí, Mexico

Cesar Marcial Escobedo-Bonilla
Departamento de Acuacultura
Instituto Politécnico Nacional–
 CIIDIR Sinaloa
Guasave, Mexico

Ángel Josué Félix-Sastré
Departamento de Biotecnología y
 Ciencias Alimentarias
Instituto Tecnológico de Sonora
Ciudad Obregón, Mexico

Juan Carlos Fernández-Macias
Coordinación para la aplicación
 de la Ciencia y la Tecnología
 (CIACYT)
Universidad Autónoma de San Luis
 Potosí
San Luis Potosí, Mexico

Jesus Guadalupe García-Clark
Programa de Maestría en Ciencias
 en Recursos Naturales
Instituto Tecnológico de Sonora
Ciudad Obregón, Mexico

Jose Reyes Gonzalez-Galaviz
CONACYT–Instituto Tecnológico de
 Sonora
Ciudad Obregón, Mexico

Ana Karen González-Palomo
Centro de Biociencias
Universidad Autónoma de San Luis
 Potosí
San Luis Potosí, Mexico

Joseph Heras
Department of Biology
California State University
San Bernardino, California

Jorge Armando Jimenez-Avalos
Unidad Guadalajara
Centro de Investigación y Asistencia
 en Tecnología y Diseño del
 Estado de Jalisco (CIATEJ)
Guadalajara, Mexico

Jorge Montiel Montoya
Departamento de Biotecnología
 Agrícola
Instituto Politécnico Nacional–
 CIIDIR Sinaloa
Guasave, Mexico

Velia Verónica Rangel-Ramírez
Centro de Biociencias
Universidad Autónoma de San Luis
 Potosí
San Luis Potosí, Mexico

Libia Zulema Rodriguez-Anaya
CONACYT–Instituto Tecnológico de
  Sonora
Ciudad Obregón, Mexico

Rodolfo Iván Valdez Vega
Centro Universitario de Ciencias de
  la Salud
Universidad de Guadalajara
Guadalajara, Mexico

Introduction

Health is an important topic that concerns us all. At present, in the midst of a human pandemic, health-related issues have acquired a renewed significance in many aspects of human life and the environment. This has a worldwide impact due to the increased risk of infectious disease outbreaks as a result of climate change, destruction of natural habitats and ever-increasing contact of human settlements with wildlife and/or exposure to polluted environments (Rohr et al., 2011). These elements are mainly a result of anthropogenic activities such as industrial development and expansion of agriculture areas. Other intrinsic components include the increasing amount of people with inadequate nutritional habits, along with the fact of increasing decline of effective drugs against microbial pathogens and/or lack of convenient disease diagnostic tools. Among the current non-communicable diseases of concern, cardiovascular disease, chronic respiratory diseases, different types of cancer and diabetes are some of the most important (WHO, 2018). Many outbreaks of infectious diseases formerly thought to be eradicated or controlled are still occurring, along with the existence of multidrug-resistant bacteria that have evolved with time (Klemm et al., 2018). The health-oriented genomic sciences have as their objective to design diagnosis and treatments (including those that are personalized) oriented towards the solution of the aforementioned health issues. Thanks to the current knowledge on molecular biology, genetics, and OMICS sciences, it is possible nowadays to conduct studies on genomes and transcriptomes, giving a wide viewpoint on genes, their interaction with environmental factors and their relationship to certain diseases. Further, infectious and non-infectious diseases also have influenced other aspects such as productivity and labor time, affecting the economy both of the producer and the consumer. The very human activities related to diseases such as agriculture and industry are also affected by the impact of non-communicable and infectious diseases. This also extends to other productive areas including aquaculture and ecology (Molster et al., 2018; Theodore et al., 2016).

Currently, transcriptome analysis using RNA-seq is a method that has revolutionized and expanded the understanding of the complexity of gene expression in eukaryotic cells, since with this technique it is possible to identify and quantify transcripts and their isoforms, the latter being a significant advantage over the other previously used techniques (EST, microarrays, qPCR) (Wang et al., 2009).

Other activities that have an important influence in human populations is access to food sources of high-quality protein content. This aspect includes animal products of aquatic origin, such as those from fisheries and aquaculture. Aquaculture is becoming a paramount animal production activity, as a source of high-quality animal protein as well as an economic activity providing employment for vulnerable sectors such as fishermen and other rural workers, and a productive enterprise for many developing countries (Escobedo-Bonilla, 2016; FAO, 2018). Despite its importance, the industry has been affected by a series of problems that threaten its development. Among these are the biological aspects related to domestication of new species that may be susceptible to culture, as well as physiological factors that may determine the viability of highly appreciated species in fisheries that may be susceptible to farming. Another essential aspect is the presence of infectious diseases in aquaculture. Transcriptomic analyses of immune and/or defense responses and tissues can help unveil the functional gene relationships within an organism in response to viral and bacterial infections and parasite infestations. All these issues can be better studied and understood in order to propose suitable and effective management and/or control strategies through the application of transcriptomics (Chandhini & Rejish-Kumar, 2018).

Regarding the immune responses between vertebrate organisms such as the human being, compared to that of invertebrates such as shrimp, it is interesting to analyze the genes and effectors present in one and the other systems. Nowadays, the technology to analyze gene expression and its relationship with other factors has greatly advanced in ways that it is now possible to quantify gene expression in a global manner (RNA-seq), in a specific cell type (scRNA), or even to measure the gene expression of each individual under the host–pathogen interaction (dual-RNA-seq) (Figure I.1).

The application of transcriptomics has reached a broad range of topics and species. In aquaculture, transcriptomics have been used to elucidate different physiological, genetic and genomic features and host–pathogen interactions of various commercially important species such as shrimp and fish. In aquatic ecology, transcriptomics have been applied to determine aspects of evolution and adaptations of fish. Transcriptomics can also be used to find the feasibility of domestication and culture potential of certain fish species of importance for fisheries. On the other hand, transcriptomics can be applied to various human health issues and infectious diseases, including cancer and diabetes, as well as infectious diseases caused by viruses, bacteria and protozoan pathogens. The genes that become activated upon any of these diseases or pathogens and their effect are important clues towards understanding the role of genes during the disease process. The comparison of gene expression between health and disease may indicate the presence of altered genes which could become biomarkers that would be used to determine the presence of a certain health condition (Figure I.1).

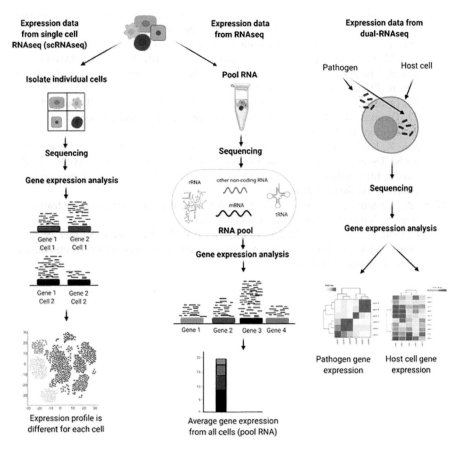

FIGURE I.1
Strategies used to analyze different types of gene expression. (Created with BioRender.com.)

References

Chandhini S, Rejish-Kumar VJ (2018). Transcriptomics in aquaculture: Current status and applications. *Reviews in Aquaculture* 11: 1379–1397.

Escobedo-Bonilla, CM (2016). Emerging infectious diseases affecting farmed shrimp in Mexico. *Austin Journal of Biotechnology and Bioengineering* 3(2): 1062–1064.

FAO (2018). *The State of World Fisheries and Aquaculture 2018. Meeting the Sustainable Development Goals.* Rome. Licence: CC BY-NC-SA 3.0 IGO: FAO.

Klemm EJ, Wong VK, Dougan G (2018). Emergence of dominant multidrug-resistant bacterial clades: Lessons from history and whole-genome sequencing. *Proceedings of the National Academy of Sciences* 115(51): 12872–12877.

Molster CM, Bowman FL, Bilkey GA, Cho AS, Burns BL, Nowak KJ, Dawkins HJ (2018). The evolution of public health genomics: Exploring its past, present, and future. *Frontiers in Public Health* 6: 247.

Rohr JR, Dobson AP, Johnson PTJ, Kilpatrick MA, Paul SH, Raffel TR, Ruiz-Moreno D, Thomas MB (2011). Frontiers in climate change-disease research. *Trends in Ecology and Evolution* 26 (6): 270–277.

Theodore K, Lalta S, La Foucade A, Cumberbatch A, Laptiste C (2016). Responding to NCDs under severe economic constraints: The links with universal health care in the Caribbean. *Economic Dimensions of Noncommunicable Diseases in Latin America and the Caribbean* 133.

Wang Z, Gerstein M, Snyder M (2009). RNA-Seq: A revolutionary tool for transcriptomics. *Nature Reviews Genetics* 10(1): 57–63.

WHO (2018). *Noncommunicable Diseases*. Geneva: WHO. Retrieved October 19, 2020, from WHO website: https://www.who.int/news-room/fact-sheets/detail/noncommunicable-diseases.

1

Shrimp Transcriptomics: Genome and Physiological Features

Cesar Marcial Escobedo-Bonilla

Instituto Politécnico Nacional–CIIDIR Sinaloa, Guasave, Mexico

CONTENTS

1.1 Introduction

Shrimp farming is an important animal production activity with over 40 years of existence (Chamberlain, 2010; FAO, 2018). Its importance relies on its great ability to expand to developing countries, contributing to improve their economy, providing an income to vulnerable populations and producing high-quality animal protein for human consumption (Pillay & Kutty, 2005; Escobedo-Bonilla, 2013). In 2016, farmed shrimp production reached 4.9 million tonnes, the most farmed species being *Litopenaeus vannamei* (86%) and *Penaeus monodon* (14%), respectively (FAO, 2018).

Since its beginning, shrimp aquaculture has sought to improve not just the culture techniques but also the shrimp species used in the industry in order to increase production. Appealing traits of farmed shrimp include ease of breeding and high spawning rate, fast growth and size, its ability to acclimate to changing environment factors, efficient feed conversion rate and pathogen resistance (Briggs et al., 2005; Pillay & Kutty, 2005; Leu et al., 2011; Yuan et al., 2018).

DOI: 10.1201/9781003212416-2

Programs aimed to produce domesticated shrimp arose in the 1970s–1980s in Tahiti and the USA, which produced shrimp lines of specific pathogen-free and specific pathogen-resistant strains of shrimp for farming purposes (Briggs et al., 2005; Wyban, 2007; Lightner, 2011). Although these programs focused on phenotypic traits such as growth rate and pathogen-free or pathogen-resistant status, such shrimp still showed performance pitfalls depending on various environmental and/or management factors (Schuur, 2003; Briggs et al., 2005; Moss et al., 2012). These were the first attempts to manipulate shrimp population genomics through phenotypic features.

With the development of molecular tools, it is possible to study the relationship between certain desired features and the genes involved with them. Various techniques have been developed in the last two decades to obtain transcriptomic information, including expressed sequence tags (EST), cDNA microarrays, suppression subtractive hybridization (SSH) and next-generation sequencing (NGS) (Robalino et al., 2007; Leelatanawit et al., 2008; Leu et al., 2011; Li et al., 2012). These tools have increased the knowledge on gene expression and the function of various aspects of shrimp life cycles, genome organization, gene function, physiology of key processes such as reproduction, sex determination, development and growth, digestion, defense function and tolerance to environmental stress. Transcriptomics will help to improve the way to regulate sex ratios in farmed and wild animal populations, expanding growth rates in shorter times and increasing disease resistance, among other traits (Santos et al., 2014; Chandhini & Rejish-Kumar, 2019).

1.2 Transcriptomic Methods

1.2.1 Expressed Sequence Tags (ESTs)

The first technique used to generate transcriptomic data from organisms with unknown genomes was the expressed sequence tags (ESTs). Here, randomly sequenced cDNA clones from a cDNA library produced short sequences (Leu et al., 2011). Further, ESTs were used to describe transcribed regions of a genome in tissues under specific conditions. This method has been an important source for gene identification and structure, novel alleles, characterization of single nucleotide polymorphism (SNP) and genome annotation (O'Leary et al., 2006; Leu et al., 2011).

The random cDNA sequencing has generated from 190 up to a maximum of 13,700 ESTs with a size ranging from 200 to 800 base pairs (bp) (Cesar et al., 2008; Leu et al., 2011). The use of ESTs has been done on various shrimp species, most of which are used in aquaculture: *Litopenaeus vannamei, L. setiferus* (Gross et al., 2001; O'Leary et al., 2006), *Penaeus monodon* (Tassanakajon et al., 2006; Leelatanawit et al., 2008), *Marsupenaeus japonicus* (Yamano & Unuma, 2006) and *Fenneropenaeus chinensis* (Xiang et al., 2008).

Studies using ESTs have produced putative gene sequences ranging from 44 to nearly 7,500 reported with this method (Table 1.1). These gene sequences are related to the following known or potential functions: defense, putative receptor, signal transduction or hormonal function, sex differentiation and reproduction, growth, cell shape, motility, extracellular matrix, DNA replication and repair, transcription/translation and ribosomal RNAs, metabolism and homeostasis, digestive functions, transport, membrane structure and channel proteins, lysosomal and proteosomal genes (Table 1.2). Other ESTs produced gene sequences different from those listed previously as well as genes of unknown function and novel sequences with no matches in public databases (Gross et al., 2001; Tassanakajon et al., 2006; Leu et al., 2011). In *P. monodon* postlarvae, analyses showed a total of 6,671 EST related to 624 genes of which 360 were unique genes in non-infected shrimp (Leu et al., 2007).

The EST technique represented a first approach to studying the shrimp transcriptome producing hundreds of novel and unique gene sequences of shrimp. This information was subsequently used by other techniques to compare gene expression in tissues/organs of animals exposed to a variety of environmental conditions, stress factors and pathogen infections, to determine the role of such genes in the response against those conditions (Carulli et al., 1998).

1.2.2 cDNA Microarrays

Later, some studies were done with the cDNA microarray technique, which compares the differential expression of genes from EST libraries. This method uses a set of cDNA clones arrayed in either nylon membranes (O'Leary et al., 2006) or on glass slides (microchip binding) at a density of 1,000 clones cm^{-2} (Carulli et al., 1998; Byers et al., 2000). Then the microarray is hybridized with mRNA from one or more cells or tissue samples which have been previously fluorescence-labeled (Cy3 or Cy5) dCTP (Carulli et al., 1998). The intensity of the hybridization is directly proportional to the expression level. This method is expensive, but it enables analysis of the expression of thousands of genes at once. One drawback of the technique is that only known genes can be screened, since these have been previously determined by EST and printed on the microchip. Hence, its use for detection of novel genes is inadequate, but it is very useful in measuring the expression of specific genes (Byers et al., 2000).

Most of the studies using cDNA microarrays have focused on the shrimp defense response against viral infection. The expression of various differentially expressed genes in shrimp species including *Litopenaeus stylirostris* (Dhar et al., 2003), *L. vannamei* (Robalino et al., 2007), *Penaeus monodon* (Lu et al., 2011) and *Fenneropenaeus chinensis* (Shi et al., 2016) have been reported upon different health and dietary conditions (Table 1.2).

TABLE 1.1

Transcriptome Methods and Tissues Used to Determine ESTs, Contigs and Genes

Method	Years	Species	Tissues	No. ESTs/ Contigs	No. Genes	References
EST	1999–2008	P. monodon	Pereon, pleopod, eyestalk	151	60	Lehnert et al. (1999)
		L. vannamei	Hepatopancreas, hemocytes	2,045	44	Gross et al. (2001)
		L. vannamei	Hemolymph, gills, hepatopancreas, lymphoid organ, eyestalk, nerve cord	13,652	7,466	O'Leary et al. (2006)
		P. monodon	Eyestalk, hepatopancreas, hematopoietic tissue, hemocytes, lymphoid organ, ovary	10,100	4,845	Tassanakajon et al. (2006)
		M. japonicus	Eyestalk	1,988	46	Yamano and Unuma (2006)
		P. monodon	Whole postlarva	6,671	624	Leu et al. (2007)
		L. vannamei	Abdominal muscle	311	160	Cesar et al. (2008)
		F. chinensis	Pereon	10,446	3,120	Xiang et al. (2008)
		L. stylirostris	Hepatopancreas	40	25	Dhar et al. (2003)
		L. vannamei	Hemocytes, gills, hepatopancreas	2,469	89	Robalino et al. (2007)
cDNA		P. monodon	Ovaries, testes	4,992	55	Karoonuthaisiri et al. (2009)
Microarray[a]	2003–2016	P. monodon	Ovaries, testes	5,568	1,439	Leelatanawit et al. (2011)
		P. monodon,	Hemocytes, gills, hepatopancreas	5,885	30	Lu et al. (2011)
		L. vannamei	Hemocytes, gills, hepatopancreas	5,885	30	Lu et al. (2011)
		F. chinensis	Muscle, hepatopancreas	59,137	1,539	Shi et al. (2016)
		M. japonicus	Hemocytes	291	77	He et al. (2004)
		P. monodon	Testes	166	77	Leelatanawit et al. (2008)
		P. monodon	Ovaries	109	93	Preechaphol et al. (2010)

		Species	Tissue	EST/config	No. genes	Reference
SSH[b]	2004–2017	L. vannamei	Hepatopancreas	179	73	Gao et al. (2012)
		P. monodon	Gut, gills	n.a.	152	Shekar et al. (2013)
		L. vannamei	Hepatopancreas	140	54	Peng et al. (2016)
		L. vannamei	Eyestalk, ovary	14,849	1,914	Ventura-López et al. (2017)
		L. vannamei	Whole larvae	882,339	109,169	Li et al. (2012)
		L. vannamei	Hemolymph, hepatopancreas	92,821	42,336	Guo et al. (2013)
		L. vannamei	Abdominal muscle, hepatopancreas, gills, pleopods	110,474	26,224	Ghaffari et al. (2014)
		M. japonicus	Animal-vegetal embryo poles	7,652	298	Sellars et al. (2015)
		F. merguiensis	Hepatopancreas, stomach, eye stalk, nerve cord, testes, ovaries, androgenic gland, muscle/epidermis	128,424	57,640	Powell et al. (2015)
RNA-seq	2012–2018	L. vannamei	Hepatopancreas	—	64,271	Chen et al. (2015)
		L. vannamei	Gills	466,293	349,012	Zhang et al. (2016)
		P. monodon	Heart, muscle, hepatopancreas, eyestalk	239,135	69,089	Nguyen et al. (2016)
		L. vannamei	Muscle, hepatopancreas	—	72,120	Dai et al. (2017)
		P. monodon	Eyestalk, stomach, ovary, testes, gills, hemolymph, hepatopancreas, lymphoid organ, muscle, whole embryo, nauplii, zoea, mysis	462,772	236,388	Huerlimann et al. (2018)
		P. monodon	Muscle	7,106,289	18,115	Yuan et al. (2018)
		M. japonicus	Muscle	5,632,117	16,734	Yuan et al. (2018)

[a] EST/config means number of unigenes used in microarray; No. genes means identified genes showing differential expression in the study.
[b] EST/config means number of cDNA clones used in the analyses.

TABLE 1.2

Some of the Genes Discovered and Their Role in Metabolism Using Transcriptome Methods

Method	Species	Genes	Putative Function	References
EST	*P. monodon*	Cytochrome oxidases I, II, III,	ATP metabolism	Leu et al. (2007)
		ATP synthase F0 subunit 6, NADH	ATP metabolism	
		Opsin	Signaling and cellular processes	
		Phosphopyruvate hydratase, zinc	Carbohydrate metabolism	
		Trypsin, Chemotrypsin BII	Proteinases and inhibitors	
		Ferritin	Metabolism/glycolysis	
		PmAV	Defense	
		Hemocyanin	Defense	
		Troponin C isotype gamma	Signal transduction	
		Arginine kinase	Amino-acid metabolism	
		Translation elongation factor EF-1α	Genetic information processing	
EST	*L. vannamei*	Penaeidin 2,3,4	Defense	O'Leary et al. (2006)
		Protease inhibitor	Proteinases and inhibitors	
		Crustin	Defense	
		Serpin	Proteinases and inhibitors	
		STAT	Signal transduction	
		Imd	Defense	
		Cu-metallothionein-2	Antioxidant activity	
		Chitinase	Carbohydrate metabolism	

Method	Species	Protein	Category	Reference
EST	*P. monodon*	Crustins	Defense	Tassanakajon et al. (2006)
		Prophenoloxidase activating factor	Defense	
		β-1,3-D-Glucan-Binding Protein	Defense	
		Protease-1	Proteinases and inhibitors	
		Protease inhibitor Kazal-type	Proteinases and inhibitors	
		Ras-like GTP-binding protein RHO	Apoptosis	
		IAP-associated or Viaf1	Apoptosis	
		Heat-shock protein A, B	Defense	
EST	*L. vannamei*	Heat-shock protein 70 KDa, 82 Kda	Defense	Gross et al. (2001)
		Transgutaminase	Defense/clotting	
		Lectin C-type	Pattern recognition receptor	
		Penaeidin 3a/3b/3c	Defense	
		Peroxinectin	Defense	
		Selenoprotein W	Genetic information processing	
		Superoxide dismutase	Transport and catabolism	
		Thioredoxin	Defense	
		Serine protease	Proteinases and inhibitors	
		Fatty acid binding protein	Lipid metabolism	
cDNA	*L. stylirostris*	Lipopolysaccharide and β-1,3-glucan binding protein (LGBP)	Defense	Dhar et al. (2003)
Microarray		Methionine adenosyltransferase	ATP metabolism	

(Continued)

TABLE 1.2 *(Continued)*

Some of the Genes Discovered and Their Role in Metabolism Using Transcriptome Methods

Method	Species	Genes	Putative Function	References
		Crustacyanin A2 subunit	Retinol binding protein	
		Dehydrogenase	ATP metabolism	
		18S and 40S ribosomal RNAs	Translation	
		CCCH-type zinc-finger	Defense	
		Integrin α, β	Cell adhesion	
		GULF adaptor protein	Apoptosis	
cDNA		Thioredoxin reductase	Oxidative stress	
Microarray	*L. vannamei*	Cathepsins A,B,D,L,	Proteases	Robalino et al. (2007)
		Tudor staphylococcal nuclease	RNA interference	
		Zn finger protein	Signal transduction	
		Interleukin enhancer binding factor 2	Transcription	
		Crustin	Defense	
		Lysozyme	Defense	
cDNA	*P. monodon*	Mo-penaeidin	Defense	Lu et al. (2011)
Microarray	*L. vannamei*	Transglutaminase	Defense/clotting	
		Kazal-type proteinase inhibitor	Proteinases and inhibitors	
		PolyI:C	Defense	
		Arginine kinase	Muscle/cytoskeleton/motility	
		BUD31 homolog	Protein synthesis	
		Clottable protein 2	Defense	

Method	Species	Gene/Protein	Function	Reference
cDNA Microarray	*F. chinensis*	Caspase 2	Signal transduction	Shi et al. (2016)
		Cathepsin C	Protease	
		Calnexin	Defense	
		HMGBb	Transcription	
		Pacifastin	Defense	
		Inhibitor of tyrosine kinase (low)	Proteinases and inhibitors	
		Neutralized protein (low)	Protein digestion and absorption	
		LTB4DH (low)	Enzymes	
		Calreticulin (low)	Cell adhesion	
		Bystin (low)	Ribosomal biogenesis	
cDNA Microarray	*P. monodon*	Cyclin B (high)	Cellular processes	Karoonuthaisiri et al. (2009)
		Receptor for activated protein kinase C (RACK) (high)	Signal transduction	
		Cortical rod protein (high)	Chromosome and associated proteins	
		Ras-related nuclear protein (Ran)	Signal transduction	
		Growth factor receptor bound protein 2 (Grb2)	Signaling pathway	
		TGF-b receptor interacting protein 1	Cellular processes	
SSH	*M. japonicus*	Interferon receptor 1-bound protein 4	Defense	He et al. (2004)
		Integrin beta 4 binding protein	Protein binding	
		Chaperonin containing TCP1 (CCT)	Chemokine receptor type 4	

(Continued)

TABLE 1.2 *(Continued)*
Some of the Genes Discovered and Their Role in Metabolism Using Transcriptome Methods

Method	Species	Genes	Putative Function	References
		PolyA binding protein	Genetic information processing	
		Synapse-associated protein SAP90/PSD95	Postsynaptic specialization	
		Arginine kinase Pen m2	Amino-acid metabolism	
		Cytochrome oxidase I	Energy metabolism	
		Elongation factor-1α	Genetic information processing	
		GTP-binding protein	Signaling and cellular processes	
SSH	*P. monodon*	26S proteasome non-ATPase subunit 12	Genetic information processing	Leelatanawit et al. (2008)
		Receptor for activated protein kinase C (RACK)	Signal transduction	
		Myelodysplasia/myeloid leukemia factor	Transcription factor	
		Ribosomal protein S6	Translation	
		Elongation factor 1α, 2	Genetic information processing	
		Calreticulin	Cell adhesion/defense	
		Ficolin	Signaling and cellular processes	
SSH	*P. monodon*	Selenophosphate synthetase	ATP binding	Preechaphol et al. (2010)
		S-adenosylmethionine synthetase	Lipid metabolism	
		T-complex protein 1 subunit epsilon	Biological process	
		Coatomer protein complex subunit beta	Cellular process	
		Peritrophin	Extracellular matrix components	

Method	Species	Gene/Protein	Function/Category	Reference
SSH	*F. chinensis*	β-tubulin	Cytoskeleton	Xie et al. (2010)
		Spermatogonial stem cell renewal factor	Cell growth and death	
		Cytochrome oxidase c subunit I	Metabolism	
		Cytochrome oxidase c subunit II	Metabolism	
		Transposase	Genetic information processing	
		Clottable protein	Defense	
		Hemocyanin	Defense	
		C-type lectin 1	Defense	
		Cathepsin C, L	Defense	
SSH	*L. vannamei*	Chitinase	Defense	Gao et al. (2012)
		Zinc proteinase mpc 1	Defense	
		Trypsin	Defense	
		Chymotrypsin 1	Proteinases and inhibitors	
		Lysozyme	Defense	
		Cathepsin B	Cellular process	
		Crustin	Defense	
		Arginine kinase	Muscle, cytoskeleton, motility	
		Na^+/K^+-ATPase a-subunit	Ion transport and osmoregulation	
SSH	*P. monodon*	Anti-lipopolysaccharide factor	Defense	Shekar et al. (2013)
		Intracellular fatty acid binding protein	Energy and metabolism	
		Ubiquitin conjugating enzyme E2	Cellular process	
		Calreticulin	Cellular process	

(Continued)

TABLE 1.2 (*Continued*)
Some of the Genes Discovered and Their Role in Metabolism Using Transcriptome Methods

Method	Species	Genes	Putative Function	References
		Tetraspanin-8	Signaling and cellular processes	
		DEAD-box helicase 5	ATP-dependent RNA helicase	
		Heat shock protein 70	Defense	
		Metallothionein	Signaling and cellular processes	
SSH	*L. vannamei*	Cytochrome b	ATP metabolism	Peng et al. (2016)
		Alpha-N-acetylgalactosaminidase-like	Transport and catabolism	
		Transcription factor X-box binding protein 1	Transcription factor	
		Trypsin	Proteinases and inhibitors	
		Renin receptor	Receptor and associated proteins	
		Cadherin EGF LAG seven-pass G-type receptor 1	Receptor and associated proteins	
		Cubilin	Receptor and associated proteins	
SSH	*L. vannamei*	Integrin alpha PS3	Receptor and associated proteins	Ventura-López et al. (2017)
		Protein fem-1 homolog A	Sex determination/differentiation	
		Protein sex-lethal	Sex determination/differentiation	
		Male-specific lethal 3 homolog	Sex determination/differentiation	
		Crustins 1-4	Defense	
		Ras GTP exchange factor K	Signaling and cellular processes	
		Argonaut-like protein	Chromosome and associated proteins	

Method	Species	Gene/Protein	Category	Reference
		Putative RNA-binding protein Luc7-like 2	Proteins: genetic information processing	
RNA-seq	*L. vannamei*	Integrin beta-like protein 1	Cellular processes	Li et al. (2012)
		DEAD box helicase homolog family member (ddx-23)	Transcription	
		Cold-inducible RNA-binding protein	Transcription	
		Thioredoxin-like 4A	Transcription	
		C-type lectin	Defense	
		Lysozyme	Defense	
RNA-seq	*L. vannamei*	HSP90	Defense	Guo et al. (2013)
		Toll-like receptors	Signal transduction	
		Caspase-3, -6	Signal transduction	
		Apaf-1	Signal transduction	
		Chitinase	Proteinases and inhibitors	
		Aminopeptidase N	Proteinases and inhibitors	
		Trypsin	Proteinases and inhibitors	
RNA-seq	*L. vannamei*	Chmotrypsin	Proteinases and inhibitors	Wei et al. (2014)
		Alpha amylase	Proteinases and inhibitors	
		Carboxipeptidase A	Proteinases and inhibitors	
		Carboxipeptidase B	Proteinases and inhibitors	
		Pancreatic triacylglycerol lipase	Proteinases and inhibitors	
		Vitellogenin	Reproductive process	

(Continued)

TABLE 1.2 *(Continued)*

Some of the Genes Discovered and Their Role in Metabolism Using Transcriptome Methods

Method	Species	Genes	Putative Function	References
RNA-seq	*F. merguiensis*	Dmc1	Sexual reproduction	Powell et al. (2015)
		Homeobox protein Nkx-2.6	Morphogenesis	
		Transcription factor SOX-8	Development maturation	
		Exostosin-2	Development growth	
		Farnesoic acid O-methyltransferase	Metabolism	
		Ecdysteroid receptor E75	Metabolism	
		Sex-lethal	Sex determination	
		Male-specific lethal-2, 3	Sex determination	
		Transformer-2a, 2c	Sex determination	
RNA-seq	*M. japonicus*	Argonaute 1	Germ line	Sellars et al. (2015)
		Germ cell-less (gcl)	Germ line	
		Gustavus (gus)	Germ line	
		Maelstrom (mael)	Germ line	
		β-catenin/armadillo	Axis formation/segmentation	
		Triacylglycerol lipase	Lipid metabolism	
		Fatty acid synthase	Lipid metabolism	
		Glycerol kinase	Lipid metabolism	
RNA-seq	*L. vannamei*	Phosphatidate phosphatase	Lipid metabolism	Chen et al. (2015)
		Long-chain acyl-CoA synthetase	Lipid metabolism	
		Long-chain-acyl-CoA dehydrogenase	Lipid metabolism	

Method	Species	Gene/Protein	Function	Reference
RNA-seq	*L. vannamei*	Acyl-CoA oxidase	Lipid metabolism	Gao et al. (2015)
		Ecdysteroid regulated-like protein	Hormone biosynthesis	
		Molting fluid carboxypeptidase	Proteinases and inhibitors	
		Ecdysone receptor	Metabolism	
		Ecdysone-induced protein 74EF	Metabolism	
		Ecdysteroid receptor E75	Metabolism	
		Ecdysone-induced protein 78	Metabolism	
		Retinoid-X receptors	Hormone signaling	
		Methyl farnesoate	Hormone biosynthesis	
		Farnesoic acid	Metabolism	
		Vitellogenin gene	Reproduction	
		Crustacean calcium-binding protein	Ion transport	
		Serine/threonine-protein kinase	Ion transport	
		Sodium- and chloride-dependent GABA transporter	Ion transport	
RNA-seq	*L. vannamei*	Glycerol-3-phosphate dehydrogenase	Energy metabolism	Zhang et al. (2016)
		Phosphoenolpyruvate carboxykinase	Energy metabolism	
		X-box-binding protein	Genetic information processing	
		Transcription factor AP-2	Genetic information processing	
		Elongation factor EF-1 gamma subunit	Genetic information processing	
		5-hydroxytryptamine (serotonin) receptor	Growth and muscle development	
		Alpha-amylase	Growth and muscle development	

(Continued)

TABLE 1.2 *(Continued)*

Some of the Genes Discovered and Their Role in Metabolism Using Transcriptome Methods

Method	Species	Genes	Putative Function	References
RNA-seq	*P. monodon*	Cyclophilin	Growth and muscle development	Nguyen et al. (2016)
		Profilin	Growth and muscle development	
		Translin-associated factor-X	Growth and muscle development	
		Crustacean hyperglycemic hormone (CHH)	Growth and muscle development	
		Molt-inhibiting hormone (MIH)	Growth and muscle development	
		Ecdysteroid	Growth and muscle development	
		Ring box protein	Growth	
		G2/mitotic-specific cyclin-B3 isoform 1	Growth	
		Wee1-like protein kinase (WEE1)	Growth	
		Aurora kinase A	Growth	
RNA-seq	*Marsupenaeus japonicus*	Mitotic spindle assembly checkpoint protein (MAD2A)	Growth	Li et al. (2017)
		DNAJ homolog subfamily C member 2 (DNAJC2)	Growth	
		Serine/threonine-protein kinase Chk1	Growth	
		Mitotic checkpoint serine/threonineprotein kinase BUB1 beta	Growth	
		Cyclin-dependent kinase inhibitor 1B	Growth	
		Cyclin E	Growth	
		Protein bric-a-brac 1	Sex differentiation	

Method	Species	Gene/Protein	Function	Reference
RNA-seq	*L. vannamei*	E3 ubiquitin-protein ligase HECTD1	Structure development	Dai et al. (2017)
		Mitogen-activated protein kinase 14	Protein binding	
		Hemolymph clottable protein	Defense/clotting	
		Spectrin alpha chain	Protein binding	
		Multiple pdz domain protein	Protein binding	
		Selenophosphate synthetase	Biological process	
		Myosin heavy chain	ATP binding	
		Vitellogenin gene	Reproduction	
		Molting fluid carboxipeptidase A	Proteinases and inhibitors	
		β-N-acetyl glucosaminidase	Transport and catabolism	
		Ecdysteroid regulated-like protein	Hormone biosynthesis	
RNA-seq	*L. vannamei*	Chitin synthase	Carbohydrate metabolism	Gao et al. (2017)
		Glucose-6-phosphate isomerase	Carbohydrate metabolism	
		Chitinase	Carbohydrate metabolism	
		Myosin	Signaling and cellular processes	
		C-type lectin	Defense	
		Anti-lipopolysaccharide factor	Defense	
		Serine protease 1	Proteinases and inhibitors	
		Spaetzle	Defense	
RNA-seq	*P. monodon*	NADH dehydrogenase subunit 2	Energy metabolism	Huerlimann et al. (2018)
		Cytochrome c oxidase subunit	ATP metabolism	
		Crustin P	Defense	
		Prophenoloxidase activating enzyme	Defense	

The cDNA microarrays have been used to study differentially expressed genes from healthy and white spot syndrome virus (WSSV)-infected shrimp, where 24 genes mostly related to defense functions were differentially expressed in WSSV-infected shrimp compared to healthy controls (Dhar et al., 2003). Immune-related genes in *L. vannamei* were differentially expressed (89 out of 2,469) under different health (WSSV early and late infection, challenged with heat-killed bacteria and/or fungal spores, treated with duck IgU dsRNA, compared to controls) and physiological conditions (WSSV-infected at 32°C vs. WSSV-infected at 27°C) (Robalino et al., 2007). It also has been used to locate the expression of defense genes in *P. monodon* and *L. vannamei* showing that 30 known genes were differentially expressed in hemocytes, gills and hepatopancreas upon immune stimulation with microbial cell wall components such as lipopolysaccharide, β-1-3 glucan, peptidoglycan and the use of Chinese medicinal herbs (Lu et al., 2011). Also, the differential expression of defense-related genes in a WSSV-resistant strain of *Fenneropenaeus chinensis* surviving a WSSV experimental challenge was evaluated. Surviving shrimp at a late stage (264–288 h post inoculation) of WSSV infection showed 1,539 differentially expressed genes compared to controls. Of these, 78% were down-regulated at the late stage of infection. Further, 151 unique genes were recorded, of which 100 were up-regulated at this stage. These genes were determined to be involved in molecular functions such as binding and catalytic activity (61%), cellular components (cell parts and organelles) (59%) and biological processes (cellular, metabolic processes and biological regulation) (61%). Surviving shrimp showed 74 known annotated genes that were specifically expressed and may be related to their WSSV-resistant status and/or be involved in the antiviral defense response (Shi et al., 2016).

Other studies have used cDNA arrays to identify reproduction and the role of genes in shrimp gonad maturation. Since reproduction and selective breeding are main traits for shrimp broodstock and source of postlarvae for pond stocking, these features are very important to determine the genes related to gonad development and maturation (Karoonuthaisiri et al., 2009; Leelatanawit et al., 2011). From previous studies where shrimp EST cDNA from testes and ovaries were obtained, cDNA microarrays were constructed with 4,992 EST from testes (3,072) and ovaries (1,920), and printed in glass slides at 21 × 15 spot squares (Karoonuthaisiri et al., 2009). Most of the known ESTs (54% testes and 41% ovaries) had gene ontology (GO) functions of "cellular components" after annotation. Other annotated ESTs had functions of "molecular function" (19% and 22%, respectively) and "biological processes" (27% and 37%, respectively) (Karoonuthaisiri et al., 2009). Differential gene expression was observed in the three function categories during ovarian development. High expression levels at early (III) and late (IV) cortical rod stages of gonad development were found in certain genes such as the cyclin B, cell division cycle 25 (Cdc25), Cdc16, mitogen-activated protein kinase binding protein 1 (MAPKBP1), receptor for activated protein kinase C (RACK), cortical rod protein, polehole, hepatocarcinogenesis-related transcription factor and the nuclear autoantigenic sperm protein (NASP). This work allowed to uncover the gene molecular mechanisms controlling the oocyte maturation process and formation of cortical rods (CRs), leading to a better understanding of the reproductive maturation of domesticated *P. monodon* (Karoonuthaisiri et al., 2009). Likewise, one study used cDNA microarrays with 5,568 genes from *P. monodon* testes and ovaries, including positive (shrimp DNA) and negative (buffer, bacteria DNA, virus DNA) controls. About 70% of the ESTs were genes with known putative functions within three categories: "biological" (42% genes), "molecular" (32% genes) and "cellular components" (26% genes). (Leelatanawit et al., 2011). Of these, 1,439 genes showed significant differential expressions. Most of these genes were from testis (30.4%) and heart (28.9%), whereas the remaining genes were from the ovary (14.8%), gill (10.7%), hemocyte (9.8%), hepatopancreas (2.6%) and intestine (1.0%). Comparative gene expression levels between juveniles and brooders showed that 576 genes were highly expressed in juveniles and only 85 genes were highly expressed in brooders (Leelatanawit et al., 2011). Also, distinct gene expression profiles were observed between wild and domesticated males. It is possible that these differences may explain the different reproductive capacity in domesticated males and may lie at the transcriptomic level (Leelatanawit et al., 2011).

1.2.3 Suppression Substractive Hybridization (SSH)

Suppression substractive hybridization (SSH) can analyze minute amounts of mRNA to study isolation of genes up-regulated in one cell type or tissue compared to another (Byers et al., 2000). Advantages of SSH over previous transcriptomic methods include the ability to detect rare mRNAs and discover novel genes (Byers et al., 2000). The SSH improves differentially expressed cDNA enrichment with only one cycle of hybridization (1,000-fold) (Byers et al., 2000) because target cDNA is selectively amplified, while an undesirable sequence is simultaneously suppressed during PCR amplification (Rusaini & Owens, 2018).

With this technique, differential gene expression has been studied in seven species of penaeid shrimp: *P. monodon, M. japonicus, L. vannamei, L. stylirostris, F. chinensis, F. merguiensis* and *F. indicus*. Libraries have been constructed from whole postlarvae, and in older shrimp from whole pereon (also known as cephalothorax) or individual organs including eyestalk, hepatopancreas, gills, testes, ovaries and circulating hemocytes (Rusaini & Owens, 2018).

The first study using SSH on shrimp determined the genes involved in defense upon challenge with heat-killed bacteria and a fungus. Here, 291 cDNA clones accounted for 77 known genes, of which 25 were related to defense response (He et al., 2004). Eight of these genes were reported for the first time in shrimp (see Table 1.2). Other studies have used SSH to uncover the genes involved in gonad maturation and to determine molecular markers on gonad development in brooders. Also, SSH has been used to determine the differential expression of genes upon gonad development and maturation (Leelatanawit et al., 2008). Here, 166 ESTs matching known genes and an additional 199 unknown genes were involved in development and maturation of testes. Of these, seven known genes had low frequencies in the libraries but were well represented in the subtracted cDNA. Likewise, another study done in the same species determined the differential expression of genes in the previtellogenesis (I) and cortical rod (III) stages in ovaries of female shrimp (Preechaphol et al., 2010) finding up to 93 differentially expressed genes between stage I and stage III.

The SSH technique was used to compare substracted cDNA libraries of ovaries from triploid and diploid shrimp *Fenneropenaeus chinensis* (Xie et al., 2010). Substractions were done both forward (cDNA from triploid ovary was used as tester and cDNA from diploid ovary was used as driver) and reverse (cDNA from diploid ovary was used as tester, and cDNA from triploid ovary was used as driver) in order to determine up-regulated genes in each type of ovary. In the triploid ovary, 576 cDNA fragments assembled into 43 contigs and 70 singlets, of which 39% of the assembled sequences had no significant similarity to any known genes or proteins in the consulted databases, 13% had significant similarities to genes encoding hypothetical proteins without any annotation and the remaining 48% of sequences encoded 54 putative genes, most of them new for the species. The genes from the triploid ovary were related to metabolism (17%), extracellular matrix components (15%), genetic information processing (5%), cell growth and death (4%), immunity (1.9%), cytoskeleton (1.7%) and signal transduction/transport (1.6%). The most abundant transcripts were peritrophin, cytochrome c oxidase subunit I, cytochrome c oxidase subunit II and spermatogonial stem cell renewal factor (Xie et al., 2010).

In the diploid ovary, 376 sequenced cDNA fragments produced 25 contigs and 25 singlets. Of these, 35% had no similarity to any known genes or proteins, 11% had significant similarities with genes encoding hypothetical proteins without annotation and the remaining sequences encoded 6 putative genes. Of the 376 cDNA sequences, 25.5% were related to extracellular matrix components, followed by cytoskeleton (15%), cell growth and death (6%), genetic information processing (4%), metabolism (2.4%) and signal transduction/transport (0.5%). The most abundant transcripts were β-tubulin, thrombospondin, peritrophin, and cellular apoptosis susceptibility protein. Only two of the putative novel genes were identified to already identified genes in *F. chinensis* (Xie et al., 2010).

Genes related to metabolism and immunity were relatively higher in the forward library, whereas genes related to extracellular matrix components and cytoskeleton had significantly higher proportion in the reverse library. The triploid ovary apparently showed that up-regulated genes, such as spermatogonial stem cell renewal factor, clotting protein, antimicrobial peptide, transposase, cytochrome oxidase c subunit I and cytochrome oxidase c subunit II, were apparently up-regulated in the ovary of triploid shrimp; while genes from the reverse-subtracted library, such as cellular apoptosis susceptibility protein, tubulin, farnesoic acid O-methyltransferase, thrombospondin and heat shock protein 90, were up-regulated in the ovary of diploid shrimp. These results indicated the involvement of these genes in the ovary development of *F. chinensis* (Xie et al., 2010).

Another study determined the role of genes involved in regulation of gonad maturation in *L. vannamei* from subadult and adult female ovaries and from juvenile, subadult and adult female eyestalk. This was done to identify candidate genes controlling the maturation process (Ventura-López et al., 2017). Total RNA from eyestalk was obtained from each of the tissues and pooled for each developmental stage (juveniles, subadults and adults). Likewise, ovary RNA was obtained from subadults and adults. The cDNA libraries were constructed from these samples and used for SSH. A total of 191,500 raw reads from both cDNA-full length and SSH libraries were produced, which after trimming gave a total of 156,563 high quality reads from ovary (92,897) and eyestalk (63,666). The assembled sequences produced 14,849 contigs with an N50 of 360. Sequence annotations for the 3,300 contigs showed GO annotations of 1,914 transcripts. The GO annotations included biological processes, metabolic process, cellular process, cellular component organization or biogenesis and localization, while reproduction had only 25 transcripts. The obtained genes showed an important number of genes previously unknown for *L. vannamei*. Only 9.3% of the genes have been previously reported in the Uniref90 database for *Litopenaeus vannamei*. Thus, the transcriptomic information produced from eyestalk and ovary showed new genes associated with female shrimp reproduction (Ventura-López et al., 2017).

Other studies done with SSH have focused on the effect of low salinity on *P. monodon* (Shekar et al., 2013) and *L. vannamei* (Gao et al., 2012) as well as on low temperature in *L. vannamei* (Peng et al., 2016), respectively.

Shrimp *L. vannamei* postlarvae (0.27 ± 0.1 g) were raised in salinities of 2 or 30 g/l for 56 days and then total RNA was obtained from hepatopancreas to construct forward and reverse substractive cDNA libraries. Randomly selected clones (n = 200) were sequenced (80 forward and 120 reverse) at both 5' and 3' ends, yielding 179 high-quality ESTs (64 forward and 115 reverse). The ESTs from the forward library assembled into 26 consensus sequences including 19 singletons and 7 contigs, whereas the ESTs from the reverse library were 47 consensus sequences, 35 singletons and 12 contigs (Gao et al., 2012). Most of the forward (69.23%) and reverse (78.72%) libraries' unigenes showed significant homologies to known protein sequences in Genbank. In contrast, only 24.66% of the unigenes had no similarities to any known protein sequence in the public database. The matched unigenes represented 17 different genes in five categories, as determined using the primary functions of their encoded proteins. Genes in the forward library showed that seven identified genes were up-regulated under hyposmotic conditions, whereas in the reverse library 37 out of 47 genes showed significant homology to protein sequences in the Genbank database. The reverse library showed that 10 identified genes were down-regulated under long-term hyposmotic conditions. Among 17 genes (7 from the forward and 10 from the reverse library), 11 encoded for immune-related proteins and enzymes (Gao et al., 2012).

A similar study done in juvenile *P. monodon* (10 - 15 g) exposed to low salinity (2 g/l) for two weeks generated four libraries from gill, gut and muscle. Forward and reverse cDNA libraries were done from pooled gut or gill tissues from six animals collected from low (3 g/l) or control shrimp at normal (28 g/l) salinity, to produce two forward and two reverse SSH cDNA libraries (Shekar et al., 2013). The obtained ESTs generated 14 contigs and 38 singletons from the forward SSH cDNA library and 1 contig and 23 singletons from the reverse SSH cDNA library constructed from gill tissues. In contrast, 18 contigs and 12 singletons from the forward SSH cDNA library and 9 contigs and 39 singletons were generated from the reverse SSH cDNA library constructed from gut tissues. Significantly up-regulated genes in gill tissues included those with functions in defense, ion transport and osmoregulation, cellular processes, energy and metabolism (see Table 3.2). Many of these functions are important for osmoregulation upon salinity stress.

The effect of low temperature stress on gene expression of *L. vannamei* was determined, as this species is also cultured in temperate regions (Peng et al., 2016). Juvenile shrimp (10–15 g) were acclimated to cold temperature (13°C) for 36 h post acclimation to extract total RNA. A total of 384 clones were chosen. Of these, a total of 92 forward and 48 reverse EST libraries with highest differential expression were selected and sequenced. The assembled ESTs gave 37 contigs from the forward library and 17 from the reverse. Among the 54 examined contigs, 10 had no significant homology to any previously identified genes, 3 were homologous to unclassified genes or genes with unknown functions and 41 had known gene homologies. These ESTs were related to stress response or cell defense functional annotations, which are associated with stress resistance, along with many novel genes identified (Peng et al., 2016).

1.2.4 RNA-Seq

The first record of RNA sequencing (RNA-seq) assessed in shrimp is the transcriptome analysis of whole larvae of *Litopenaeus vannamei* (Li et al., 2012). The transcriptome contained ≈2.4 Giga bases (Gb) from which 109,169 unigenes (≈396 bp) were assembled. A total of 73,505 high-quality unigene sequences (≥200 bp) were used for annotation analysis, showing that 37.80% of such unigenes matched the NCBI Non-redundant (nr) database, 37.3% matched in Swiss-Prot, and 44.1% matched in TrEMBL. The BLAST and BLAST2Go analyses showed that 11,153 unigenes belong to 25 clusters of orthologous groups of protein (COG) categories, 8,171 unigenes were assigned into 51 GO functional groups, and 18,154 unigenes were divided into 220 Kyoto Encyclopedia of Genes and Genomes (KEGG) pathways. The transcriptome analysis using NGS techniques enhanced the information of *L. vannamei* genes, which improved the understanding of the crustacean genomics fostering these types of studies on shrimp and crustaceans (Li et al., 2012).

Since then, other works have dealt with transcriptome analyses and genome annotation of different shrimp species such as the improved transcriptome annotation of *Litopenaeus vannamei* (Ghaffari et al., 2014), the transcriptome profile of *Marsupenaeus japonicus* embryos (Sellars et al., 2015), the transcriptome analysis of the banana shrimp *Fenneropenaeus merguiensis* (Powell et al., 2015), the transcriptome annotation of *P. monodon* (Huerliman et al., 2018) and genome sequencing and annotation of *Marsupenaeus japonicus* and *P. monodon* (Yuan et al., 2018).

The transcriptome of *L. vannamei* was produced from abdominal muscle, hepatopancreas, gills and pleopods from one male to get a pooled transcriptome assembly. A high-quality *de novo* transcriptome assembly was produced with 399,056,712 trimmed reads resulting in 110,474 contigs of which 2,701 had N50 quality. Many of the sequences were related to proteins involved in "housekeeping" but some tissue-specific proteins were in the top 20 most frequent, such as Down's syndrome cell adhesion molecule (Dscam). As many as 4,493 genes with immune function were found (Ghaffari et al., 2014).

The expression of digestive enzyme genes during different larval stages (zygote, blastula, gastrula, limb bud embryo, larva in membrane, nauplius, zoea, mysis and postlarva 1) of Pacific white shrimp *Litopenaeus vannamei* was evaluated using RNA-seq (Wei et al., 2014). Total filtered reads obtained for different larval development stages (embryos, nauplii, zoea, mysis and postlarva) were 51,568,556; 52,824,674; 53,430,302; 53,902,786; 51,574,056, respectively. The independent assembly of reads produced a total of 66,815 unigenes. Of these, 32,398 had matches in the NCBI nr protein database, 29,022 in Swiss-Prot database and 26,257 were associated with 255 pathways by KEGG pathway mapping. Hundreds of unigenes were categorized within pathways related to food digestion and absorption such as salivary and/or pancreatic secretions, protein digestion and absorption. Many unigenes were annotated as digestive enzymes with essential functions during food digestion. Unigenes recognized as digestive enzymes were clustered into three major groups. The first group was the carbohydrases, comprised by three types of polysaccharidases and two types of disaccharidases. Of these, chitinase showed the highest unigene frequency (n = 67), including types 1–6 and Chid1. Alpha amylase, a fundamental polysaccharidase involved in glycogen catabolism, also was highly represented (16 unigenes). Isomaltase, a previously unrecognized enzyme in shrimp, was reported in seven unigenes. The two types of disaccharidases were identified as maltases: alpha glucosidase and beta galactosidase. The second group was peptidases, composed by three types of endopeptidases, three types of exopeptidases and one dipeptidase. The endopeptidases were trypsin, chymotrypsin and elastase, all members of the serine protease family. No aspartic digestive proteases such as pepsin or chymosin were identified. The exopeptidases included carboxypeptidase A, carboxypeptidase B and aminopeptidase N. The latter enzyme was the most frequent unigene (n = 67), followed by trypsin (n = 29). The third group was lipases composed of triacylglycerol lipases. These included gastric triacylglycerol lipase, bile salt-activated lipase and pancreatic triacylglycerol lipase, having the latter as the highest unigene occurrence (n = 13). The number of expressed enzyme unigenes increased with development. Trypsin, chymotrypsin and lipase appeared at embryo stage, whereas the expression of alpha amylase began until nauplius. The expression of this enzyme sharply increased from nauplius to zoea. Further, expression of trypsin and chymotrypsin occurred from embryo to nauplius and from nauplius to zoea. In the case of lipase, some unigenes were highly expressed at embryo, whereas others expressed until nauplius stage. The wide enzyme diversity confirmed the ability of *L. vannamei* to use different types of diets to fulfill its nutritional requirements. Also, the dynamic enzyme expression patterns during early development indicates the importance of transcriptional regulation to adapt to the diversity of diets available throughout its larval development (Wei et al., 2014).

The pooled RNA isolation and cDNA synthesis from *M. japonicus* animal and vegetal half-embryos were used for transcriptome analyses. The cDNA from 500 multiple-spawning pooled animal and vegetal half-embryos yielded 560,516 and 493,703 reads, respectively. Reads from each library were assembled and gene ontogeny analysis produced 3,479 annotated animal contigs and 4,173 annotated vegetal contigs, with 159/139 hits for developmental processes in the animal/vegetal contigs, respectively. Genes found included those for sex determination, axis determination/segmentation genes and cell-cycle regulators (Sellars et al., 2015).

Tissues (hepatopancreas, stomach, eye stalk, nerve cord, testes, ovaries, androgenic gland and muscle/epidermis) from juvenile and adult shrimp *F. merguiensis* were collected to extract total RNA and synthesize cDNA libraries. A total of 128,424 contigs were obtained which putatively contained 57,640 gene encoding proteins with a mean length of 167 amino acids (Powell et al., 2015). The most representative genes belong to the signal transduction mechanisms, general function and the transcription category, respectively.

Transcriptomic analyses were done in adult Pacific white shrimp *Litopenaeus vannamei* to determine changes in gene expression during the molting process. Whole individuals were used for RNA-seq studies in the different stages of the complete molting cycle: inter-molt (C), pre-molt (D0, D1, D2, D3, D4) and post-molt (P1 and P2) (Gao et al., 2015). Normalized cDNA libraries (n = 16) obtained from all molting stages gave a total of

148,250,000 filtered reads (7.4 Gb) with 98.7% being high-quality reads (Q20) for expression analysis. Of the total reads, between 83.1% and 88.4% matched the reference genome. The clean reads were assembled into 93,756 unigenes of which 15,582 matched known proteins in the NCBI nr database, 3,689 matched putative homologues in the nt database, 6,493 were annotated in KO, 12,873 in Swiss-Prot and 20,784 found putative homologues in the Pfam database. Per molting stage, the number of expressed genes were 55,639 (C), 56,655 (D0), 60,115 (D1), 59,636 (D2), 50,693 (D3), 57,689 (D4), 52,813 (P1) and 58,890 (P2), respectively. Of these, stage-specific expressed genes were: 3,995 (C), 4,096 (D0), 5,994 (D1), 6,664 (D2), 4,187 (D3), 5,254 (D4), 4,360 (P1) and 5,888 (P2), respectively. In contrast, 28,508 unigenes were expressed at all eight stages. Many of these genes had low variance or low expression levels, and they were considered to be either housekeeping genes or rarely expressed during molting (Gao et al., 2015). A total of 5,117 genes were differentially expressed (DEG) between any two adjacent stages (C–D0, D0–D1, D1–D2, D2–D3, D3–D4, D4–P1, P1–P2, P2–C). Of these, 4,015 DEGs occurred between D4–P1 and P1–P2, respectively, being the stages with the largest amount of DEGs. The DEGs were classified into 47 functional GO groups from three main categories: biological processes, 4,885 (42.5%); cellular components, 3,606 (31.4%); and molecular functions, 3,010 (26.1%). Immune genes (n = 236) were categorized in 36 families between molting stages. Most of them were overexpressed in stages D3–D4. During molting (D4–P1), some genes were up-regulated, including crustin and anti-lipopolysaccharide factor (ALF) which increased their normalized expression after molting 7.6- and 3.3-fold, respectively. Several unigenes of hemocyanin, chitinase, serine protease, trypsin and chymotrypsin were identified during molting. Hemocyanin expression was up-regulated at C and again at D3–D4 stages and down-regulated during D1–D3, and at P1 it was expressed at a minimum; whereas chitinase genes were expressed during D1 and D4. Serine protease, trypsin and chymotrypsin were up-regulated at C–D0, D3–D4 and P1–P2 (Gao et al., 2015).

The gene expression profile showed that most down-regulated genes occurred in molting stages D4–P1, and they were related to shrimp metabolism including cellular, metabolic, binding, catalytic and organic substance metabolic processes. Whereas in P1–P2, several genes were up-regulated, including protein digestion and absorption, neuroactive ligand-receptor interaction and lysosome-related pathways. These findings were consistent with the morphological and molecular changes observed between molting and post-molt. This study provided a basic understanding of the molecular changes happening during molting (Gao et al., 2015).

Another work characterized the gene expression profile of *L. vannamei* during its larval development until adult and throughout all the molting stages related to exoskeleton development and dynamics (Gao et al., 2017). A total of 117,539 unigenes were obtained, with half of the total assembly length

(N50) of 1,327 bp and 83,002,524 bp. The functional annotation of unigenes showed that 32,398 (48.5%) had putative homologues in the nr database, 19,363 (29.0%) had putative homologues in the nt database, and 29,022 (43.4%) had putative homologues in the Swiss-Prot database. The GO analysis associated 158,366 terms to all unigenes. The secondary GO classification sorted all unigenes within 49 functional groups belonging to the three main categories: biological processes (19,455 or 34.1%), molecular function (21,510 or 37.7%) and cellular components (16,122 or 28.2%). The full-length coding sequences (CDS) of 48,425 unigenes (41.2% of all unigenes found) were predicted, of which 19,336 had matches in the nr and Swiss-Prot databases. The remaining 29,089 CDS with no specific matches were predicted by ESTScan.

The complete exoskeleton development process was obtained in 17 sequential growth stages from zygote to adult in *L. vannamei*, including nine successive growth stages and eight molting stages by the digital gene expression profiling technique. Gene expression during all these stages showed that they gradually increased in early development. Some genes were specifically expressed in different molting stages, whereas a set of 19,809 genes showed significant differential expression between two consecutive developmental stages.

The genes involved in exoskeleton formation, development and reconstruction include (i) cuticle proteins (cuticle protein, chitin binding protein, structural constituents of cuticle), (ii) molting hormone (molting-inhibiting hormone, crustacean hyperglycemic hormone), (iii) exoskeleton degradation (chitinase, beta-N-acetyl-glucosaminidase, acetyl-hexosaminidase), (iv) mineral absorption (solute carrier family, ferritin heavy chain etc.), (v) mineral reabsorption (sodium/ potassium-transporting ATPase, clathrin heavy chain, etc.), (vi) exoskeleton synthesis (chitin synthetase, Glutamine:Fructose-6-phosphate aminotransferase), (vii) ion channel (calcium channel, potassium channel, chloride channel), (viii) molting signaling pathway (calmodulin, nitric oxide synthase, molting defective family member) and (ix) late molting genes (actin/tubulin/myosin, hemocyanin). In total, 603 unigenes were identified as being involved in exoskeleton formation, development and reconstruction. A hierarchical cluster analysis generated a global view of the gene expression pattern related to exoskeleton development, indicating that the majority of genes were up-regulated during the middle (larva in membrane/nauplius) and late (zoea/mysis) phases of early development stages. This study showed a distinct expression time and profile of exoskeleton developmental genes in the early growth stages of *L. vannamei*. The data suggested that exoskeleton formation may initiate in the larva in the membrane/nauplius stages (phase II). Gene expression related to exoskeleton synthesis, degradation, regulation, calcification and hardening were mainly expressed in zoea/mysis/postlarva stages (phase III), indicating a gradually advanced system for exoskeleton development in this phase (Gao et al., 2017).

The whole transcriptome of Kuruma shrimp *Marsupenaeus japonicus* embryos at gastrula stage was analyzed to identify growth-related genes and the molecular mechanism of early embryonic development (Li et al., 2017). A total of 108,320,302 trimmed reads accounted for 10.8 Gb and with GC content of 46.69% were assembled into 67,183 contigs (range from 201 to 14,938 bp, with mean length 1,216 bp and N50 = 2,405 bp). Of these, 46,661 unigenes were obtained with 910 bp mean size and N50 = 1,735 bp. Functional annotation showed that 13,265 unigenes (28.4% of the total) matched in NCBI nr protein sequences and 13,218 sequences were allocated into open reading frames and coding sequences. An additional 15,333 unigene coding sequences (32.86%) were successfully predicted. Classification of 14,399 unigenes using GO database assigned them to three main categories (biological process, molecular function and cellular component) and 49 functional subgroups. The biological process was classified into 21 subgroups, of which the top three were cellular (20.9%), metabolic (17.9%) and single-organism (12.2%) processes. The molecular function category showed the three top subgroups as binding (45.5%), catalytic activity (32.9%) and transporter activity (7.4%). The cellular component category included two top groups: cell and cell part, each with 19.6% of the unigenes. Other groups included organelle (11.6%) and membrane (10.5%). The GO and KEGG unigenes classification identified 37 growth-related genes. These genes showed broad expression and abundance variations. The most expressed genes were the ring box protein and the G2/mitotic-specific cyclin-B3 isoform 1, indicating their importance in cell proliferation and growth during the embryonic development of the Kuruma shrimp (Li et al., 2017).

Similarly, *P. monodon* brooders were used for transcriptomic analyses using nine different tissue types (eyestalk, stomach, testes, ovaries, gills, hemolymph, hepatopancreas, lymphoid organ and muscle). Also, four larval stages (embryo, nauplii, zoea and mysis) were used to collect pooled replicate tissues for RNA extraction and cDNA synthesis for transcriptomic analyses (Huerlimann et al., 2018). The combined sequencing results gave a total of 462,772 contigs after software filtering. The final assembly consisted of 236,388 transcripts with an assembly size of 226 megabases. Annotation against the non-redundant arthropod (nrA) database successfully matched 62,679 contigs, of which 48,456 had a GO mapping and 25,201 were completely annotated. The top-hit species was the freshwater amphipod *Hyalella azteca* with over 20,000 hits, followed by *P. monodon* with over 2,500 hits. Other penaeid shrimp species included *Litopenaeus vannamei*, *Marsupenaeus japonicus* and *Fenneropenaeus chinensis*, within the 12 highest represented species. The *P. monodon* transcriptome analyses showed genes involved in eukaryotic cellular processes such as RNA replication, transcription, biological processes, metabolism and signaling (Huerlimann et al., 2018).

Other studies used NGS tools to annotate the genome of two shrimp species *M. japonicus* and *P. monodon*. A total of 5,632,117 contigs and 80,444 unigenes accounting for 1.94 Gb were found for *M. japonicus* and 7,106,289 contigs, 89,473 unigenes for a total 2.04 Gb for *P. monodon*, respectively

(Yuan et al., 2018). A total of 16,734 genes were annotated for *M. japonicus* and 18,115 genes for *P. monodon*. Of these, 4,845 candidate orthologs were found between two genomes. Among these gene sequences, horizontal gene transfer (HGT) between viruses and shrimp was investigated, but no genes were found. Only 13 genes of *M. japonicus* and 15 genes of *P. monodon* showed moderate homology to WSSV genes (21%–63% identities). Likewise, horizontal gene transfer was examined between shrimp genomes and 16 candidate HGTs were found between these two shrimp genomes. Of these, 13 were shared by two shrimp genomes except for Ankp, CTC, and mtkA. Other 14 HGTs were identified in *L. vannamei*, eight genes in *M. japonicus* and nine genes in *P. monodon*, which indicated these genes may be transferred from a bacterial source to the ancestors of penaeid shrimp (Yuan et al., 2018).

More recently, the complete genome sequence of the Pacific white shrimp *L. vannamei* has been published using NGS. The genome is 1.66 Gb long. The scaffold N50 was 605.56 kb, with 25,596 protein-coding genes and a high proportion of simple sequence repeats (SSR) (>23.93%). The GC content was 35.7%, the content of transposable elements was 16.2% and the number of predicted noncoding RNA genes was 2,777. The genome covered > 94% of the unigenes assembled from RNA-Seq data and 94.8% of the conserved core eukaryotic genes. The k-mer analysis indicated that repetitive sequences accounted for ~78% of the genome, far more than those identified in the final assembly (49.38%), indicating that some repetitive sequences were still missing. (Zhang et al., 2019).

The genes in *L. vannamei* had longer exons (259 bp) and more exons per gene (5.9), compared to crustaceans *Daphnia pulex* and *Parhyale hawaiensis*. The *L. vannamei* genome showed 762 expanded gene families and 16,291 species-specific genes (>57% of all the genes), including genes related to myosin complex, chitin binding, metabolic process and signaling transduction. Orphan genes can contribute to lineage-specific adaptation. A total of 3,369 orphan genes were identified. These displayed fewer exons (4.5 exons/gene), but longer (292 bp/gene), in contrast to the average length of all genes (5.9 exons and 260 bp per gene). Gene features such as special gene structure characters and special temporal/spatial expression patterns, suggest independent *de novo* origins. Such orphan genes displayed transcriptome unigenes present in other penaeid shrimp such as *Fenneropenaeus chinensis* (83.1%) and *P. monodon* (64.6%), but rarely observed in other decapods like *Exopalamon carincauda* (1.95%), *Eriocheir sinensis* (1.1%), and *Neocaridina denticulata* (3.8%), suggesting that these orphan genes are lineage-specific, and may contribute to penaeid-specific adaptation (Zhang et al., 2019).

Most of the SSRs were located in intergenic regions (24.6%) and introns of protein-coding genes (22.1%), being less frequent in exons (1.4%). A telomere component (AACCT)n, identified by fluorescence *in situ* hybridization, is longer than many other SSRs. The longest SSR found was 13,769 bp in size. The length of (AACCT)n located in introns is longer than those found in other genomic regions. The DNA transposons (9.33%) and long interspersed elements (LINEs, 2.82%) comprised the two major classes of transposable elements (Zhang et al., 2019).

The NGS tools have also been used to determine the differential gene expression upon nitrite stress (Guo et al., 2013), different lipid sources at various salinities (Chen et al., 2015), diverse salinity conditions (Zhang et al., 2016), expression of growth-related genes (Nguyen et al., 2016) and different feed-efficiency shrimp batches (Dai et al., 2017).

The transcriptome information of shrimp exposed to nitrite was done to determine the molecular mechanisms underlying the nitrite exposure and to identify genes involved in apoptosis, immune defense and detoxification. Juvenile shrimp *L. vannamei* (4.4 ± 1.8 g) were exposed to 20 mg/l nitrite concentration in water at 5 g/l salinity at different times (12, 24, 48 and 72 h) (Guo et al., 2013). A total of 58.79 million raw reads were obtained which were reduced to 52.89 million clean reads containing 4.76 Gb. These data were used in the *de novo* assembly to get 92,821 contigs with 26.32 million bases assembled into 42,336 unigenes. These comprised a coverage of 73% of the NCBI transcriptome library (73,505 transcriptome sequences) (accession: PRJNA73443), whereas 27% of the obtained unigenes were not found in the NCBI library. A total of 10,777 (25.46%) GO functions were obtained of which 49% were biological processes, 33% were cellular components and 18% molecular functions. A total of 2,445 unigenes had functions of cell killing, death, growth and defense system (Guo et al., 2013).

The transcriptome differential response of *L. vannamei* hepatopancreas as a function of dietary lipid source at low salinity to reveal the key pathways and genes sensitive to the change of dietary lipid sources was studied (Chen et al., 2015). Juvenile shrimp (1.86 ± 0.32 g) were acclimated to a salinity of 3 g/l salinity prior to the experiment. Shrimp were acclimated to experimental conditions by being fed three times daily with a commercial diet (40% crude protein, 8% crude lipid, 12% ash, 30% carbohydrates, 16.7 kJ g/1 digestible energy), and during the experiment, the shrimp were fed three times daily with three diets containing fatty acids of different origins (beef tallow [BT], fish oil [FO] and equal combinations of soybean oil + BT + linseed oil [SBL]) at the same concentrations. At the end of the experiment, the shrimp were fasted for 24 h before hepatopancreas sampling (n = 5) for transcriptome analyses. A total of 135.5 million high-quality reads were obtained which included 26,034 genes and 38,237 isogenes. The clusters of orthologous groups (COG) prediction functionally classified the genes into 25 categories, in which most were in general functions, transcription and signal transduction. Relevant gene expressions in the pathways of fatty acid elongation and unsaturated fatty acids biosynthesis indicates that *L. vannamei* may possess the ability to synthesize DHA and EPA from α-linolenic acids at low salinity (Chen et al., 2015).

A similar study used transcriptomic analysis of gills in *L. vannamei* to determine the osmotic response and osmoregulation at three salinities in order to produce a reference transcriptome of gill tissue upon different salinities (Zhang et al., 2016). Juvenile shrimp (3–4 cm) were acclimated at salinity 30 g/l, and 24°C for one week before the experiments. Then, the shrimp were randomly placed in three different salinity conditions: marine (30 g/l) brackish (15 g/l) and freshwater (0.5 g/l) at 24°C. Total RNA was purified from five shrimp per treatment and cDNA was obtained from mRNA in two replicates per group. A total of 124,914,870; 119,250,450, and 105,487,350 raw reads were generated from the marine, brackish and freshwater groups, respectively. The *de novo* assembly was done with 349,652,670 high-quality reads, which produced 466,293 transcripts. Assembly analysis created 349,012 unigenes. Of these, 24,554 (7.04%) showed significant BLASTx matches in the nr database and 16,826 (4.82%) showed significant matches in the Swiss-Prot database. The top five species showing BLASTx hits were *Daphnia pulex* (2,214 unigenes; 7.4%), *Tribolium castaneum* (1,587 unigenes; 5.5%), *Pediculus humanus* (1,114 unigenes; 3.6%), *Branchiostoma floridae* (595 unigenes; 2.5%), and *Nasonia vitripennis* (637 unigenes, 2.2%). Among the 16,646 unigenes annotated in KOG, 675 were annotated and classified into 26 KOG categories. The general functions and signal transduction mechanisms were the highest represented, followed by translation, ribosomal structure and biogenesis (Zhang et al., 2016). Most of the genes related to ion transport and amino acid metabolism were up-regulated, proving that ion channels and free amino acids were important osmotic effectors in *L. vannamei*. Likewise, energy-related genes were activated in freshwater versus marine salinity, suggesting an increased energy demand of osmoregulation and rebalancing of energy homeostasis to promote energy status and guarantee the normal physiological activities of *L. vannamei* (Zhang et al., 2016).

The *de novo* assembly of *P. monodon* transcriptome was done with mRNA from heart, muscle, hepatopancreas and eyestalk tissues (Nguyen et al., 2016). These tissues produced 78,575,346 raw paired-end reads which after trimming yielded 62,327,726 (79.32%) clean reads which assembled into 69,089 unigenes with an average size of 448 bp. These unigenes were annotated against public protein databases followed by nr-NCBI, Swiss-Prot, GO, COG, and KEGG classification and identified growth-related genes. A total of 12,321 unigenes (17.83%) were annotated to the five databases, whereas the remaining 56,768 unigenes did not have a significant match to any known protein in the public databases. The non-annotated unigenes may be novel transcribed sequences specific to *P. monodon*, or may have been too short to allow for statistically meaningful matches. The GO predicted functions of the assembled unigenes revealed that 5,686 (8.23%) sequences were included into 58 functional groups such as biological, cellular (3,363, 59.15%), metabolic (3,264, 57.40%) processes, biological regulation (1,597, 28.09%), pigmentation (1,429, 25.13%), and localization (1,012, 17.80%). Of the total genes annotated in this study, 165 had a putative function in growth and muscle development. Differential expression analysis showed growth- and development-related unigenes which were expressed in *P. monodon* in a tissue-specific manner (Nguyen et al., 2016).

A study was done on the molecular basis of feed efficiency through residual feed intake of shrimp. Feeding efficiency is essential to developing programs on genetic breeding in *L. vannamei* (Dai et al., 2017). Here, sequencing of 12 samples from muscle and one hepatopancreas from a high-efficiency group (HFH), a low-efficiency group (LFL) and a control group (MG) originating from two families was done.

The sequencing of these 12 samples generated 680,960,100 raw reads, with an average read length of 150 bp. The filtered reads were assembled *de novo* into 72,120 unigenes with lengths between 201 and 38,364 bp and an average length of 1,484 bp and median size (N50) of 2,841 bp. Among these unigenes, 47,509 genes had a length more than 500 bp and 29,446 had a length more than 1,000 bp. The total length of the unigenes was up to 107,047,713 bp.

The 72,120 unigenes were submitted to different databases (BLAST nr protein, Swiss-Prot, GO and KEGG), to obtain gene sequence annotation for gene identity. Most of the genes were annotated in the nr database (32.9%), followed by the Swiss-Prot (27.6%) and GO (25.6%) databases. The least annotated genes (10.7%) were found in the KEGG database. Of these, 7,742 (10.7%) unigenes were assigned to 271 different pathways. Only 7,014 unigenes (9.7%) were annotated in all of the four databases.

Differences in gene expression patterns between the efficient and inefficient individuals within each family were found. According to the residual feed intake (RFI) distribution and family backgrounds, 10 individuals were divided into a high-efficiency group (HG: HFH1, HFH2 and HFH3), a low-efficiency group (LG: LFL2, LFL3 and LFH3) and a medium control group (MG: LFH1, LFH2, HFL1 and HFL2). No significant differences were found in average daily gain between these groups (P > 0.05).

In the HG group, 56 genes were differentially expressed (DEGs) as compared with MG (q value < 0.05), of which 27 were down-regulated and 29 were up-regulated. Almost all (55/56) of these genes were up-regulated > 5 fold between HG and MG. A relatively large number of genes (n = 348) were differentially expressed between LG and MG (q value < 0.05), of which 61 were down-regulated and 287 were up-regulated. Also here, up-regulated genes were expressed > 5 fold or more. A total of 383 nr genes between the two sets of DEGs were considered to be associated with RFI. These genes were mostly involved in cell proliferation, growth and signaling, glucose homeostasis, energy and nutrient metabolism (Dai et al., 2017).

1.3 Perspectives

Transcriptome analyses on cultured crustaceans and shrimp have been used for 20 years, and now it has become an important tool to study different aspects relevant to crustacean aquaculture, including reproduction and breeding, larviculture, growth, nutrition and defense. The evolution of technologies used to determine the transcriptome has been fast, and these various techniques have been used to increase the knowledge on the gene transcript composition in the different species and to determine their expression in various processes at different development stages and environmental conditions.

With the advent of NGS technologies such as RNA-seq, the capabilities to map and measure the amount of transcribed genes and their expression patterns (up- or down-regulation) within a process or life stage will be greatly improved. This technology is expected to promote research in order to increase the knowledge on genomics, gene expression and quantification, physiology, defense and their relationship with the environment and other external factors.

It is expected that NGS technology will foster the knowledge on various aspects of molecular biology such as genetic resources, analysis of molecular pathways, differential and correlated gene expression and identification of molecular markers. These research areas will help expand our knowledge on the presently cultured species, improving their genetic makeup, husbandry and breeding. It also will help expand their phenotypic features according to certain environmental factors in a particular region. Other species that are appealing to aquaculture would also benefit from transcriptomic studies by developing strategies to their culture. It also will give insights on their genomic organization as well as information on their metabolic, environmental and physiological requirements in order to be successfully domesticated.

References

Briggs M, Funge-Smith S, Subasinghe RP, Phillips M (2005) *Introductions and movement of two penaeid shrimp species in Asia and the Pacific*. Rome, Italy: Food and Agriculture Organization of the United Nations.

Byers RJ, Hoyland JA, Dixon J, Freemont AJ (2000) Subtractive hybridization—genetic takeaways and the search for meaning. *International Journal of Experimental Pathology* 81: 391–404.

Carulli JP, Artinger M, Swain P, Root CD, Chee L, Tulig C, Guerin J, Osborne M, Stein G, Lian J, Lomedico PT (1998) High throughput analysis of differential gene expression. *Journal of Cellular Biochemistry* 72 (Supplements 30/31): 286–296.

Cesar JR, Zhao B, Yang J (2008) Analysis of expressed sequence tags from abdominal muscle cDNA library of the Pacific white shrimp *Litopenaeus vannamei*. *Animal* 2(9): 1377–1383.

Chamberlain GW (2010) History of shrimp farming. In: *The shrimp book*. V Alday-Sanz editor. Chicago, USA. Nottingham University Press, pp. 1–34.

Chandhini S, Rejish-Kumar VJ (2019) Transcriptomics in aquaculture: current status and applications. *Reviews in Aquaculture* 11: 1379–1397.

Chen K, Li E, Xu Z, Li T, Xu C, Qin JG, Chen L (2015) Comparative transcriptome analysis in the hepatopancreas tissue of Pacific white shrimp *Litopenaeus vannamei* fed different lipid sources at low salinity. *PLoS ONE* 10(12): e0144889.

Dai P, Luan S, Lu X, Luo K, Kong J (2017) Comparative transcriptome analysis of the Pacific white shrimp (*Litopenaeus vannamei*) muscle reveals the molecular basis of residual feed intake. *Scientific Reports* 7: 10483.

Dhar AK, Dettori A, Roux MM, Klimpel KR, Read B (2003) Identification of differentially expressed genes in shrimp (*Penaeus stylirostris*) infected with white spot syndrome virus by cDNA microarrays. *Archives of Virology* 148: 2381–2396.

Escobedo-Bonilla CM (2013) Application of RNA Interference (RNAi) against viral infections in shrimp: a review. *Journal of Antivirals and Antiretrovirals* 5(3): 1–12.

FAO (2018) *The State of World Fisheries and Aquaculture 2018. Meeting the sustainable development goals*. Rome. Licence: CC BY-NC-SA 3.0 IGO: FAO.

Gao W, Tan B, Mai K, Chi S, Liu H, Dong X, Yang Q (2012) Profiling of differentially expressed genes in hepatopancreas of white shrimp (*Litopenaeus vannamei*) exposed to long-term low salinity stress. *Aquaculture* 364–365: 186–191.

Gao Y, Wei J, Yuan J, Zhang X, Li F, Xiang J (2017) Transcriptome analysis on the exoskeleton formation in early developmental stages and reconstruction scenario in growth-moulting in *Litopenaeus vannamei*. *Scientific Reports* 7(1): article 1098 https://doi.org/10.1038/s41598-017-01220-6.

Gao Y, Zhang X, Wei J, Sun X, Yuan J, Li F, Xiang J (2015) Whole transcriptome analysis provides insights into molecular mechanisms for molting in *Litopenaeus vannamei*. *PLoS ONE* 10(12): e0144350.

Ghaffari N, Sanchez-Flores A, Doan R, Garcia-Orozco KD, Chen PL, Ochoa-Leyva A, Lopez-Zavala AA, Carrasco JS, Hong C, Brieba LG, Rudiño-Piñera E, Blood PD, Sawyer JE, Johnson CD, Dindot SV, Sotelo-Mundo RR, Criscitiello MF (2014) Novel transcriptome assembly and improved annotation of the whiteleg shrimp (*Litopenaeus vannamei*), a dominant crustacean in global seafood mariculture. *Scientific Reports* 4: 7081.

Gross PS, Bartlett TC, Browdy CL, Chapman RW, Warr GW (2001) Immune gene discovery by expressed sequence tag analysis of hemocytes and hepatopancreas in the Pacific white shrimp, *Litopenaeus vannamei*, and the Atlantic white shrimp, *L. setiferus. Developmental and Comparative Immunology* 25: 565–577.

Guo H, Ye CX, Wang AL, Xian JA, Liao SA, Miao YT, Zhang SP (2013) Transcriptome analysis of the Pacific white shrimp *Litopenaeus vannamei* exposed to nitrite by RNA-seq. *Fish & Shellfish Immunology* 35: 2008–2016.

He N, Liu H, Xu X (2004) Identification of genes involved in the response of haemocytes of *Penaeus japonicus* by suppression subtractive hybridization (SSH) following microbial challenge. *Fish & Shellfish Immunology* 17: 121–128.

Huerlimann R, Wade NM, Gordon L, Montenegro JD, Goodall J, Sean M, Tinning M, Siemering K, Giardina E, Donovan D, Sellars MJ, Cowley JA, Condon K, Coman GJ, Khatkar MS, Raadsma HW, Maes GE, Zenger KR, Jerry DR (2018) De novo assembly, characterization, functional annotation and expression patterns of the black tiger shrimp (*Penaeus monodon*) transcriptome. *Scientific Reports* 8: 13553.

Karoonuthaisiri N, Sittikankeawa K, Preechaphol R, Kalachikov S, Wongsurawat T, Uawisetwathana U, Russo JJ, Ju J, Klinbunga S, Kirtikara K (2009) ReproArrayGTS: a cDNA microarray for identification of reproduction-related genes in the giant tiger shrimp *Penaeus monodon* and characterization of a novel nuclear autoantigenic sperm protein (NASP) gene. *Comparative Biochemistry and Physiology, Part D* 4: 90–99.

Leelatanawit R, Klinbunga S, Aoki T, Hirono I, Valyasevi R, Menasveta P (2008) Suppression subtractive hybridization (SSH) for isolation and characterization of genes related to testicular development in the giant tiger shrimp *Penaeus monodon*. *BMB Reports* 41(11): 796–802.

Leelatanawit R, Uawisetwathana U, Klinbunga S, Karoonuthaisiri N (2011) A cDNA microarray, UniShrimpChip, for identification of genes relevant to testicular development in the black tiger shrimp (*Penaeus monodon*). *BMC Molecular Biology* 12: 15.

Lehnert SA, Wilson KJ, Byrne K, Moore SS (1999) Tissue-specific expressed sequence tags from the black tiger shrimp *Penaeus monodon*. *Marine Biotechnology* 1(5): 465–476.

Leu JH, Chang CC, Wu JL, Hsu CW, Hirono I, Aoki T, Juan HF, Lo CF, Kou GH, Huang HC (2007) Comparative analysis of differentially expressed genes in normal and white spot syndrome virus infected *Penaeus monodon*. *BMC Genomics* 8: 120.

Leu JH, Chen SH, Wang YB, Chen YC, Su SY, Lin CY, Ho JM, Lo CF (2011) A review of the major penaeid shrimp EST studies and the construction of a shrimp transcriptome database based on the ESTs from four penaeid shrimp. *Marine Biotechnology* 13: 608–621.

Li C, Weng S, Chen Y, Yu X, Lu L, Zhang H, He J, Xu X (2012) Analysis of *Litopenaeus vannamei* transcriptome using the next-generation DNA sequencing technique. *PLoS One* 7(10): e47442.

Li P, Wang Y, Wang Y, Guo H (2017) Whole transcriptome sequencing and analysis of gastrula embryos of *Marsupenaeus Japonicus* (Bate, 1888). *Madridge Journal of Aquaculture Research & Development* 1(1): 1–7.

Lightner DV (2011) *Status of shrimp diseases and advances in shrimp health management.* In: *Proceedings of the seventh symposium on diseases in Asian aquaculture. Fish health section.* MG Bondad-Reantaso, B Jones, F Corsin and T Aoki editors. Selangor, Malaysia: Asian Fisheries Society, pp. 121–134.

Lu JK, Hsiao YT, Wu JL (2011) Applications of shrimp immune DNA microarray in aquaculture. In: *Diseases in Asian Aquaculture VII.* MG Bondad-Reantaso, JB Jones, F Corsin and T Aoki editors. Selangor, Malaysia: Fish Health Section, Asian Fisheries Society, pp. 241–252.

Moss SM, Moss DR, Arce SM, Lightner DV, Lotz JM (2012) The role of selective breeding and biosecurity in the prevention of disease in penaeid shrimp aquaculture. *Journal of Invertebrate Pathology* 110: 247–250.

Nguyen C, Nguyen TG, Nguyen LV, Pham HQ, Nguyen TH, Pham HT, Nguyen HT, Ha TT, Dau TH, Vu HT, Nguyen DD, Nguyen NTT, Nguyen NH, Quyen DV, Chu HH, Dinh KD (2016) De novo assembly and transcriptome characterization of major growth-related genes in various tissues of *Penaeus monodon*. *Aquaculture* 464: 545–553.

O'Leary NA, Trent III HF, Robalino J, Peck MET, McKillen DJ, Gross PS (2006) Analysis of multiple tissue-specific cDNA libraries from the Pacific white-leg shrimp, *Litopenaeus vannamei*. *Integrative and Comparative Biology* 46(6): 931–939.

Peng J, Wei P, Chen X, Zeng D, Chen X (2016) Identification of cold responsive genes in Pacific white shrimp (*Litopenaeus vannamei*) by suppression subtractive hybridization. *Gene* 575(2, Part 3): 667–674.

Pillay TVR, Kutty MN (2005) *Aquaculture. Principles and practices*. 2nd ed. Oxford, UK: Blackwell Publishing Co.

Powell D, Knibb W, Remilton C, Elizur A (2015) *De-novo* transcriptome analysis of the banana shrimp (*Fenneropenaeus merguiensis*) and identification of genes associated with reproduction and development. *Marine Genomics* 22: 71–78.

Preechaphol R, Klinbunga S, Khamnamtong B, Menasveta P (2010) Isolation and characterization of genes functionally involved in ovarian development of the giant tiger shrimp *Penaeus monodon* by suppression subtractive hybridization (SSH). *Genetics and Molecular Biology* 33(4): 676–685.

Robalino J, Almeida JS, McKillen D, Colglazier J, Trent III HF, Chen YA, Peck MET, Browdy CL, Chapman RW, Warr G, Gross PS (2007) Insights into the immune transcriptome of the shrimp *Litopenaeus vannamei*: tissue-specific expression profiles and transcriptomic responses to immune challenge. *Physiological Genomics* 29: 44–56.

Rusaini, Owens L (2018) Suppression subtractive hybridization in penaeid prawns: an approach in identifying diseases and differentially expressed genes. *Journal of Aquaculture & Marine Biology* 7(1): 00177. doi: 10.15406/jamb.2018.07.00177.

Santos AC, Blanck DV, De Freitas P (2014) RNA-seq as a powerful tool for penaeid shrimp genetic progress. *Frontiers in Genetics* 5: 298.

Schuur AM (2003) Evaluation of biosecurity applications for intensive shrimp farming. *Aquacultural Engineering* 28 (1–2): 3–20.

Sellars MJ, Trewin C, McWilliams SM, Glaves RSE, Hertzler PL (2015) Transcriptome profiles of *Penaeus (Marsupenaeus) japonicus* animal and vegetal half-embryos: identification of sex determination, germ line, mesoderm, and other developmental genes. *Marine Biotechnology* 17: 252–265.

Shekar MS, Kiruthika J, Ponniah AG (2013) Identification and expression analysis of differentially expressed genes from shrimp (*Penaeus monodon*) in response to low salinity stress. *Fish & Shellfish Immunology* 35: 1957–1968.

Shi X, Kong J, Meng X, Luan S, Luo K, Cao B, Liu N, Lu X, Deng K, Cao J, Zhang Y, Zhang H, Li X (2016) Comparative microarray profile of the hepatopancreas in the response of "Huanghai No. 2" *Fenneropenaeus chinensis* to white spot syndrome virus. *Fish & Shellfish Immunology* 58(2016): 210–219.

Tassanakajon A, Klinbunga S, Paunglarp N, Rimphanitchayakit V, Udonmkit A, Jitrapakdee S, Sritunyalucksana K, Phongdara A, Pongsomboon S, Supungul P, Tang S, Kuphanumart K, Pichyangkura R, Lursinsap C (2006) *Penaeus monodon* gene discovery project: the generation of an EST collection and establishment of a database. *Gene* 384: 104–112.

Ventura-López C, Galindo-Torres PE, Arcos FG, Galindo-Sánchez C, Racotta IS, Escobedo-Fregoso C, Llera-Herrera R, Ibarra AM (2017) Transcriptomic information from Pacific white shrimp (*Litopenaeus vannamei*) ovary and eyestalk, and expression patterns for genes putatively involved in the reproductive process. *General and Comparative Endocrinology* 246: 164–182.

Wei J, Zhang X, Yu Y, Li F, Xiang J (2014) RNA-Seq reveals the dynamic and diverse features of digestive enzymes during early development of Pacific white shrimp *Litopenaeus vannamei*. *Comparative Biochemistry and Physiology, Part D* 11: 37–44.

Wyban J (2007) Domestication of Pacific white shrimp revolutionizes aquaculture. *Global Aquaculture Advocate* (July/August): 42–44.

Xiang J, Wang B, Li F, Liu B, Zhou Y, Tong W (2008) *Generation and Analysis of 10,443 ESTs from cephalothorax of Fenneropenaeus chinensis. 2nd International Conference on Bioinformatics and Biomedical Engineering,* Shanghai, China, 74–80.

Xie Y, Li F, Wang B, Li S, Wang D, Jiang H, Zhang C, Yu K, Xiang J (2010) Screening of genes related to ovary development in Chinese shrimp *Fenneropenaeus chinensis* by suppression subtractive hybridization. *Comparative Biochemistry and Physiology Part D: Genomics and Proteomics* 5(2): 98–104.

Yamano K, Unuma T (2006) Expressed sequence tags from eyestalk of kuruma prawn, *Marsupenaeus japonicus. Comparative Biochemistry and Physiology A. Molecular and Integrative Physiology* 143(2): 155–161.

Yuan R, Hu Z, Liu J, Zhang J (2018) Genetic parameters for growth-related traits and survival in Pacific white shrimp, *Litopenaeus vannamei* under conditions of high ammonia-N concentrations. *Turkish Journal of Fisheries and Aquatic Sciences* 18: 37–47.

Zhang D, Wang F, Dong S, Lu Y (2016) *De novo* assembly and transcriptome analysis of osmoregulation in *Litopenaeus vannamei* under three cultivated conditions with different salinities. *Gene* 578: 185–193.

Zhang X, Yuan J, Sun Y, Li S, Gao Y, Yu Y, Liu C, Wang Q, Lv X, Zhang X, Ma KY, Wang X, Lin W, Wang L, Zhu X, Zhang C, Zhang J, Jin S, Yu K, Kong J, Xu P, Cheng J, Zhang H, Sorgeloos P, Sagi A, Alcivar-Warren A, Liu Z, Wang L, Ruan J, Chu KH, Liu B, Li F, Xiang J (2019) Penaeid shrimp genome provides insights into benthic adaptation and frequent molting. *Nature* 10 (356): 1–14.

2

Transcriptomics as a Mechanism to Study Crustacean Host–Pathogen Interactions

Jose Reyes Gonzalez-Galaviz
CONACYT–Instituto Tecnológico de Sonora, Ciudad Obregón, Mexico

Jesus Guadalupe García-Clark
Instituto Tecnológico de Sonora, Ciudad Obregón, Mexico

Cesar Marcial Escobedo-Bonilla
Instituto Politécnico Nacional, Guasave, Mexico

Libia Zulema Rodriguez-Anaya
CONACYT–Instituto Tecnológico de Sonora, Ciudad Obregón, Mexico

CONTENTS

2.1 Introduction

Capture fisheries and aquaculture are activities of great importance for food production, representing high economic revenues in international markets, with a total amount of 339.6 million metric tons of aquatic products in 2015–2016. Of these, 156.1 million metric tons were produced from aquaculture and the rest from capture fisheries (FAO, 2018). At present, aquaculture represents one of the most important marine production activities due to its high throughput and profitability under controlled conditions. Among the aquaculture marine products, three main groups stand out: finfish, mollusks and crustaceans. In 2016, crustaceans accounted for an annual production value of 7.87 million metric tons. The main cultured species is the Pacific white shrimp (*Litopenaeus vannamei*) with 4.16 million metric tons. Other species of economic importance are shrimp *Penaeus monodon*, *Marsupenaeus japonicus* and *Fenneropenaeus chinensis*; the crayfish *Procambarus clarkii*; the freshwater prawns *Macrobrachium nipponense* and *Macrobrachium rosenbergii*; the Chinese mitten crab *Eriocheir sinensis*, among others (FAO, 2018; Rao et al., 2016; Zeng et al., 2013). Pathogens have been responsible for massive mortalities up to 100% in crustacean production, causing significant economic losses. Hence, studies are needed to determine the causal agents of mortalities and their relationship with environmental factors, in order to propose methods and strategies to reduce or to avoid negative impacts to production, economy, environmental pollution problems and, in extreme cases, the disappearance of crustacean populations (Sookruksawong et al., 2013; Veloso et al., 2011).

Transcriptomics is a novel technique that allows to study the host–pathogen interaction from a gene expression viewpoint, since the host cell and pathogen functions are dependent on gene activation and expression. Therefore, transcriptomics permits to determine gene expression patterns and even to observe its changes when cellular stressors are induced. In addition, the importance of transcriptomics for the analysis of expression, mapping and annotation of genes in crustaceans involves the characterization and identification of genes related to various functions such as growth and immune response (Arockiaraj et al., 2011; Pongsomboon et al., 2008; Rao et al., 2016; Rodriguez-Anaya et al., 2018; Zhong et al., 2017).

Due to the increasing interest of behavior and functionality of nucleic acids and their derivatives, techniques have been developed to foster the generation of knowledge through genomics and transcriptomics. However, in recent years, procedures have gone from simple extractions to complete sequencing of genomes, transcriptomes and proteomes in a wide variety of organisms. Transcriptomic studies may be one of the best ways to analyze genetic behavior of organisms. Genes have different expression patterns in response to environmental or physiological changes. Animals constantly express genes related to specific metabolic activities (Horgan & Kenny, 2011; Picard et al., 2015; Suravajhala et al., 2016).

Relative expression analysis is a conventional methodology that measures the ratio of gene expression gradients between a gene of interest and a steady-expressed (housekeeping) gene, under experimental conditions (Pfaffl, 2001). This way, the expression of the gene of interest can be measured and quantified in order to expand the knowledge on physiological activities in organisms. Nevertheless, it requires knowledge about the gene sequences to be studied, becoming totally dependent on sequencing techniques and other studies to understand their functionality. This makes a detailed transcriptome study of any organism almost impossible (Page & Stromberg, 2011). Novel methodologies have been developed, such as *de novo* assembly, which allows the expansion of transcriptomic knowledge, by doing deep studies on the sequences and gene expression through quantitative and qualitative methodologies in any organism (Suravajhala et al., 2016).

2.2 SNPs and Generation of Molecular Markers to Identify Resistant Families

One of the main aims of transcriptomic studies is genetic identification using molecular markers, which are important tools to identify the structure, quality and location of the labeled genes. The identification of single nucleotide polymorphisms (SNPs) within a gene of interest has been the most important way for the generation of molecular markers (Picard et al., 2015; Santos et al., 2018a). These allow the development of information libraries in which these SNPs are annexed to identify a particular genetic variant within a species. Thanks to these, it is possible to classify organisms based on their genetic variability, at either the biological, geographical or molecular level. This identification allows the generation of databases where all the markers identified and characterized in a specific organism are registered. Alternatively, their absence is recorded in certain host–pathogen interactions (Picard et al., 2015).

As an example of this, Santos et al. (2018b) analyzed the SNPs in transcripts of *L. vannamei* infected with white spot syndrome virus (WSSV). Here, 2,300 genes were expressed in growth conditions and 3,807 genes were expressed upon WSSV exposure. Within these genes, a total of 7,164 and 16,346 SNPs were identified, respectively. Such genes are involved in 275 metabolic pathways of the Kyoto Encyclopedia of Genes and Genomes (KEGG). Likewise, some of these genes were related to the inhibition of apoptosis, hemocyanin, crustin, crustacyanin and phagocytosis-inducing lectins with some variants. It is possible that some of such genes presented some immunological advantages, which may be used for the generation of molecular markers for the selection of WSSV-resistant families.

Genomic databases allow us to broaden our knowledge on genetic makeup of organisms, to write down relevant data about the genome and behavior of organisms upon changes of some biological, physical or chemical conditions on their environment. Such databases have been generated by sequencing genome fragments; despite that, with this method it was not possible to obtain complete genome sequences (Chai et al., 2013). Genomes, transcriptomes and proteomes have not been fully characterized and represent a basis for future research, where the search for the functionality or genetic relationship between species can be done (Hijikata et al., 2007). The description of these functions is mainly based on changes at the transcriptional level, caused by alterations on natural growing conditions of the organisms.

It appears that pathogen-resistant genetic lines of shrimp are increasing in popularity and use in aquaculture facilities in many countries, as they represent a means to counteract the risk of infectious diseases in a pond or farm and lower economic losses for producers. In the case of *L. vannamei*, transcriptomic studies have been done that allowed the identification of resistant lines and the possibility of determining molecular markers. Nonetheless, it has not been possible to establish genomic markers for genetic resistance to infectious diseases due to the complex nature of the shrimp genome, which has multiple repeat sequences that for a long time have prevented the correct assembly of the sequencing products (Yu et al., 2015). The variability in gene expression between a genetic line of *L. vannamei* and its interaction with a certain pathogen has been observed. As an example, the transcriptome of shrimp larvae from a genetic line showed a total of 73,505 unigenes: 11,153 unigenes were found as orthologous proteins, 8,171 unigenes were located in the gene ontology (GO) database and the other 18,154 unigenes were found in the KEGG database and were assigned to 220 metabolic pathways (Li et al., 2012). Later, Yu et al. (2014) made a comparison between the transcriptome reported by Li et al. (2012), and another study they conducted. They found 96,040 SNPs; 5,242 non-synonymous and 29,129 synonymous were predicted, with a frequency of one SNP every 476 bp. This information is very important for the design of biomarkers, since non-synonymous SNPs, contrary to synonymous, are those that can change the genetic code and type of coding amino acid in the alleles (Chu and Wei, 2019). This indicates that the number of alleles is quite diverse, potentially modifying transcripts functionality and opening the possibility of generating genetic lines with their own characteristics.

2.3 Immune Response on the Host–Pathogen Interaction Studies

Emerging diseases have been a problem in crustacean aquaculture. They are responsible for high economic losses in shrimp production, falling between 30% and 70% and up to 100% depending on the pathogen. The most damaging pathogen in shellfish farming is WSSV, which causes up to 100% mortality. The WSSV has been studied in commercial shrimp species such as *Litopenaeus vannamei*, *Marsupenaeus japonicus* and *Fenneropenaeus chinensis* (Huang et al., 2012; Li et al., 2013; Xue et al., 2013). Another viral pathogen for crustacean aquaculture is Taura Syndrome Virus (TSV). This infectious agent has caused high mortalities (40%–90%) to farmed and wild penaeid shrimp. One of its main features is the ability to be carried by disease survivor shrimp (Sookruksawong et al., 2013; Veloso et al., 2011; Zeng et al., 2013). Through the study of host–pathogen interactions, new tools can be generated to enable the prevention or correction of diseases caused by pathogens. Also, novel pathogen detection tools and molecular instruments have been developed for characterizing shrimp families and their biological features to be applied in aquaculture practices (Figueroa-Pizano et al., 2014; Rao et al., 2015; Suravajhala et al., 2016).

The severity of a disease in certain organisms depends not only on the pathogenicity of an infectious agent, but also on other factors related to the host such as susceptibility to a pathogen or its tolerance/resistance to it. Other factors such as physiological conditions of the host, presence of environmental stressors and variations of environmental conditions that may exist in the host habitat are also influential on the occurrence of a disease (Chen et al., 2016; Li et al., 2017; Xue et al., 2013). The interaction of these factors within a host can have an influence at different levels: organs, tissues, cells and/or molecular. With transcriptomics it is possible to analyze accurately, quickly and safely the transcriptomic response of the host system upon contact with pathogens, through gene expression analyses (Aoki et al., 2011; Shekhar & Ponniah, 2015). It is necessary to get a natural reference status as a control (natural conditions) and compare it with the expression data in the presence of pathogens. In this way it is possible to identify differentially expressed genes (DEGs) that may be up-regulated or down-regulated (Figure 2.1). The type and quantity of the genes affected can serve as a tool to identify the organism response mechanism upon a pathogen infection (Zhong et al., 2017).

2.3.1 Bacteria

2.3.1.1 Vibrio parahaemolyticus

Vibrio parahaemolyticus is a Gram-negative bacterium responsible for vibriosis outbreaks. Some isolates contain a plasmid (pVP70) harboring binary toxin genes (vpPirA/B), which are the causal agents of the early mortality syndrome (EMS) now known as acute hepatopancreas necrosis disease (AHPND). Clinical signs of disease include damage mainly in the hepatopancreas through liquefactive necrosis of tissues (Junprung et al., 2017). Also, Rao et al. (2015), observed changes in gene expression during infection with

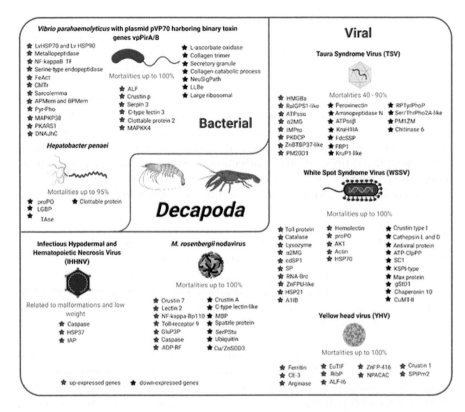

FIGURE 2.1
Main expression of genes in Decapoda upon infection of their major pathogens. (Created with BioRender.com.)

V. parahaemolyticus in *M. rosenbergii*. Their results showed 14,569 unigenes aberrantly expressed, where 11,446 genes were up-regulated and 3,103 down-regulated. The former ones were usually related to immune response mechanisms of the host. In contrast, down-regulated genes were related to the inhibition of gene expression by the presence of repressors released mainly by some pathogens. Those genes fell into the categories of metabolic pathways, regulation of actin cytoskeleton, spliceosome, RNA transport, focal adhesion, biotin metabolism, phenylalanine, tyrosine and tryptophan biosynthesis, vitamin B6 metabolism, lipoic acid metabolism and thiamine metabolism. Likewise, the most representative up-regulated unigenes were anti-lipopolysaccharide factor as an important peptide to bind to and to neutralize lipopolysaccharides available on the cellular membrane of some pathogens (Ren et al., 2012). Crustin plays an important role as antimicrobial with activity against both Gram-positive and Gram-negative bacterial strains (Arockiaraj et al., 2013a). The importance of NF-kappa B inhibitor alpha involves adaptive and innate immune responses activation and apoptosis inhibition (Arockiaraj et al., 2012a). Other molecules are also involved in defense. Transglutaminase reduces blood loss from damaged tissue and prevents the entry of pathogens into the prawn body (Arockiaraj et al., 2013b). The hemolectin and mannose-binding protein recognizes pathogens by binding to molecular structures that are specific of a given component from a pathogen (Wang & Wang, 2013). The serpin serine protease inhibitor and serpin B allow peptidase regulation, and they are recognized as the best mechanisms (Gatto et al., 2013). Chitinase plays an important role when a pathogenic invasion and immunosuppression are taking place in vertebrates and invertebrates (Ravichandran et al., 2018). Innexin 3 forms gap junction and non-junction channels that participate in cellular communication of various physiological processes (Güiza et al., 2018). Beclin 1 has a central role in autophagy (Kang et al., 2011). The heat shock proteins 21, 60, 70 and 90, arginine kinase 1 and caspase have as their main activities stress and thermal regulators in response to various environmental conditions, temporal and spatial damping of ATP and apoptosis regulation, respectively (Arockiaraj et al., 2011; Chaurasia et al., 2016; Wu et al., 2014). Also, some candidates of the down-regulated genes were perlucin-like protein and tachylectin as protein recognition patterns; cathepsin D, kazal-type serine proteinase inhibitor 4 as proteinase inhibitors; Toll pathway genes as Toll receptor 2; the MAPK signaling pathway as max protein; and other immune related genes as ferritin, peritrophin, selenoprotein W, metallothionein I, crustacyanin-like lipocalin and adenosine deaminase. Nonetheless, the latter does not describe the operation and host–pathogen interaction in its entirety, since these genes may have different functions, may be related to and be responsible for alterations in the transcriptome. These results will allow a deeper study of genes that are directly related with the presence of AHPND or vibriosis in cultured crustaceans.

As has been observed, some constant down-regulated genes in the crustacean–*V. parahaemolyticus* interaction are proteinase inhibitors, recognition proteins and the Toll-like pathway, which functions mainly as germline encoded pattern recognition receptors (PRRs) (Brubaker et al., 2015). These genes are important in detecting a wide range of pathogens, and their inhibition would cause an increase in contagion rates by microbial agents in crustacean cultures. For example, Toll-like receptors have been previously discussed by Habib and Zhang (2020) and argued about their function as receptors that examine and stimulate extracellular, transmembrane and intercellular responses by presence of pathogens, as anti-inflammatory response. Also, they are involved in the development and regulation of antimicrobial peptides of crustaceans. In the same way, a fundamental and far-from-ruled-out characteristic of Toll-like receptors is that they allow the recognition of the outer membrane protein U in *V. parahaemolyticus* (VpOmpU), which has been described as crucial in its pathogenic response in macrophages and monocytes. They also participate in MAPK activation in those cell types (Gulati et al., 2019). Additionally, two pathogenicity islands have been reported in *V. parahaemolyticus*; one of these has a type III secretion system and the effector protein VopA belonging to the YopJ effectors family. VopA has been studied as a mediator in the inhibition of the MAPK signaling pathway due to its acetyltransferase activity in kinases inhibition (Ma & Ma, 2016; Trosky et al., 2004).

For all the above, the large number of genes involved in the crustacean–*V. parahaemolyticus* interaction and its visualization at the metabolic level leave evidence of the aforementioned bacterial symptomatology. Hence, transcriptomic responses are observed in crustaceans against pathogens, but these do not rule out the potential risk to crustaceans of the activation of such responses.

Studies in these topics may find the mechanisms for culture protection or those to generate crustacean families resistant to this pathogen, generating molecular markers that may allow the identification of genes in *M. rosenbergii* related to AHPND infection.

A particular case explains the importance of the heat shock proteins LvHSP70 and LvHSP90 as a defense mechanism of *L. vannamei* against *V. parahaemolyticus* showing survivals higher than 50%. These results suggested that the presence of these proteins play an important role in countering the effects of this pathogen (Junprung et al., 2017). Further, Qin et al. (2018) exposed the transcriptome of *L. vannamei* under *V. parahaemolyticus* infection showing 2,258 DEGs, 1,017 up-regulated genes including in female pregnancy (FeAct), NF-kappaB transcription factor activity, chloride transmembrane transport (ChlTr), sarcolemma, basal plasma membrane (BPMem) and the apical plasma membrane (APMem) subclasses. Also, metallopeptidase activity, serine-type endopeptidase activity and pyridoxal phosphate binding (Pyr-Pho) had positive regulation. Likewise, 1,241 genes were considered down-regulated, where neuropeptide signaling pathway (NeuSigPath), larval locomotory behavior (LLBe), collagen catabolic process, the secretory granule, collagen trimer, large ribosomal and L-ascorbate oxidase activity subclasses had the highest enrichment scores. In contrast, Maralit et al. (2018) published information about *L. vannamei* against *V. parahaemolyticus*, where data from 3 to 6 hpi showed 815 DEGs. The up-regulated genes were analyzed, reporting anti-lipopolysaccharide factor (ALF), crustin p, serpin 3, C-type lectin 3, clottable protein 2, mitogen-activated protein kinase kinase 4 (MAPKK4), P38 mitogen-activated protein kinase (MAPKP38), protein kinase A regulatory subunit 1 (PKARS1) and DNAJ homolog subfamily C member 1-like (DNAJhC) as the most representative genes.

2.3.1.2 Hepatobacter penaei

Hepatobacter penaei is an intracytoplasmic pleomorphic Gram-negative bacterium responsible for the necrotizing hepatopancreatitis in *Litopenaeus vannamei* and is related to economic losses between 20% and 40% of the production. It is characterized by attacking only hepatopancreatic cells (Campos-García et al., 2019; Nunan et al., 2013). During infection, this bacterium causes reduction (up to 60%) in expression of genes involved in the immune response such as prophenoloxidase system, clottable protein, lipopolysaccharide and β-glucan binding protein (LGBP) and transglutaminase (TAse) in shrimp *L. vannamei*. Only a few studies exist involving this bacterium and the genes related to the host immune response; thus, this area remains an interesting subject for research (Figueroa-Pizano et al., 2014). To our knowledge there is no information about *L. vannamei* transcriptome against *Hepatobacter penaei*, and as a consequence, this is an area of research development since it is possible to generate novel data to describe this host–pathogen interaction. Likewise, this topic will allow us to understand how this pathogen enters the host and causes imbalances in its defense response, leading to disease and mortality and translating this into significant economic losses.

In contrast to what was described in the case of *V. parahaemolyticus*, the lack of transcriptomic data makes it difficult to analyze pathogenicity mechanisms in host–pathogen interaction. Despite that *H. penaei* is a pathogen of little risk to human health, this microorganism represents an important issue to farmed shrimp since it has the potential to become a widespread pathogen. This microbe has caused massive mortalities in farmed shrimp in various countries on the American continent, and it still is present in various regions. For this reason, it is necessary to have the foundation behind this interaction, to increase the possibilities of generating inhibitors or having genetic families with high defenses against this pathogen.

2.3.2 Virus

2.3.2.1 Taura Syndrome Virus (TSV)

Transcriptomic studies have identified genes related to the infection of this virus. The TSV is a selective single stranded, positive sense, RNA virus (Lightner et al., 2012) that infects tissues of ectodermic and mesodermic origin. This pathogen has caused mortalities between 40% and 95% in farmed *L. vannamei* in the Americas and various Asian countries (Flegel, 2012).

The interaction of these genes and their expression have been reported in hemocytes and hemolymph of *L. vannamei*. Of 697 known genes; 483 were up-regulated. Of these, HMGBa, integrin and ATP synthase subunit alpha were immune regulators; the integral membrane protein was an adhesive protein; the matrix metalloprotease was a protease; alpha-2-macroglobulin was a coagulation protein; and the protein kinase domain containing protein, ras-specific guanine nucleotide-releasing factor, zinc finger and BTB domain-containing protein 37-like isoform 1 and carboxypeptidase precursor were signal transducers.

Conversely, 214 genes were down-regulated, including ATP synthase subunit beta (ATPssβ), C-type lectin and two types of fibrinogen proteins were recognized as immune regulators. Peroxinectin was an adhesive protein; two types of kruppel proteins, the putative receptor protein tyrosine phosphatase and serine/threonine-protein phosphatase 2A activator-like were signal transducers. The aminopeptidase N and protease m1 zinc metalloprotease and chitinase 6 were proteases (Sookruksawong et al., 2013). This study found that mayor metabolic pathways affected by TSV infection were pathogen/antigen recognition, coagulation, proPO pathway cascade, antioxidation and protease. The variety of changes induced in metabolic expression of this interaction may be more apparent through other studies related to proteome and metabolome.

These results indicated, at least partially, the shrimp response to TSV. For instance, shrimp type C lectin is one of the shrimp genes most affected by TSV, and it is very important in the immune response. Wei et al. (2012) identified and analyzed the individual expression patterns of two types of type C lectin (LvLectin-1 and LvLectin-2) in *L. vannamei*. A similar expression pattern of these genes was observed in the hepatopancreas, muscle, gills, hemocytes, gonads and heart. However, their expression significantly increased with the presence of a virus or bacteria. These data suggested that type C lectin expression was affected by the presence of pathogens, and its positive regulation could be due to its important role against pathogens. Nevertheless, regulation could show independent expression patterns for each type of microorganism. Recently, Zhang et al. (2019), performed an *in vitro* hemocyte immunological activity assay. They observed that a lectin type C recombinant could play an important role in PO exocytosis activity stimulation by affecting phagocytic activity and hemagglutinating activity, in addition to functioning as a precursor in activation of proPO system in *L. vannamei*. On the other hand, Bi et al. (2020) studied the immunological activity of perucline, a form of type C lectin, where they observed that *Vibrio* spp. stimulated its production. In addition, they had binding affinity to Gram-positive and Gram-negative bacteria, producing agglutination only in Gram-negative bacteria. Also, by interfering RNA they observed that gene expression related to phagocytosis and four antimicrobial peptides decreased. These works show that type C lectins play an important role in the innate immune response of *L. vannamei*. An interrelation with the other DEGs shown in the TSV–*L. vannamei* transcriptomics results could exhibit the specific expression patterns of this interaction, especially in the response of the crustacean against the virus.

Transcriptomic studies allow the qualitative and quantitative description of gene families related to the shrimp defense response that may indicate the ability of crustaceans like *L. vannamei* to persist against the impact of highly pathogenic agents (Trang et al., 2019). These characteristics can be chosen in genetic selection programs to generate families with new genetic traits or to improve the performance of commercial farmed species. Nonetheless, virus mutations are a possibility, making it necessary to have well-identified genes that may be directly associated with the presence of pathogens in cultured crustaceans.

2.3.2.2 White Spot Syndrome Virus (WSSV)

The WSSV is an important pathogen in crustacean cultures. Upon virus infection, gene expression in *M. rosenbergii* showed significant values. The 8,443 up-regulated genes included lysozyme, clip domain serine proteinase 1 (cdSP1), alpha-2-macroglobulin, serine protease (SP), Toll protein, RNA-binding region-containing protein (RNA-Brc), zinc finger protein ush-like (ZnFPU-like), catalase, heat shock protein 21 (HSP21), argonaute 1 isoform B (A1IB), hemolectin, prophenoloxidase (proPO), arginine kinase 1 (AK1) and actin, which showed the highest fold change. Conversely, 5,973 genes were down-regulated. The most representative were: crustin type I, anti-viral protein, cathepsin L and D, ATP-dependent Clp protease proteolytic (ATP-ClpPP), serine carboxypeptidase 1 (SC1), kazal-type serine proteinase inhibitor 4 (KSPI-type), max protein, glutathione S transferase D1 (gStD1), chaperonin 10 and copper-specific metallothionein CuMT-II with the lowest fold change. This indicates that WSSV has a significant impact on freshwater shrimp gene expression (Rao et al., 2016). The amount of DEGs in the interaction between WSSV and *M. rosenbergii* is unusual compared to other pathogens discussed in this chapter. Such distinctiveness involves a large number of metabolic pathways affected by the presence of this virus. Some of the affected pathways include antiviral and antimicrobial proteins, pattern recognition peptides, some proteases such as alpha-2-macroglobulin, the Toll pathway and MAPK signaling, heat shock and cell death proteins. The wide variety of pathways and effectors altered by the WSSV infection makes it difficult to control the virus, since it provokes an imbalance in the crustacean metabolism resulting in death.

It was shown that some crustaceans have a lower differential response in the transcriptome against WSSV, such as *L. vannamei*, whose transcriptome had 74 up-regulated and 37 down-regulated genes. These were mainly related to antiviral and antimicrobial proteins, proteases, protease inhibitors, hemolymph coagulation, transcriptional response regulation, cell death, cell adhesion, heat shock proteins, RNA silencing; oxidative stress; immune recognition to pathogen receptors; proPO system; signaling pathway and 17 other genes related to defense responses (Chen et al., 2013). Nonetheless, *L. vannamei* also has susceptible genotypes that have responded to a WSSV infection with 4,226 DEGs, of which 2,506 were up-regulated and 1,720 down-regulated genes. Some of the genes include C-type lectin, mannose receptor, serine protease and relevant, apoptosis relevant, ubiquitin protea-some pathway relevant, integrin, heat shock proteins, lysozyme, chitinase, crustin, calreticulin, arginine kinase and hemocyanin (Xue et al., 2013). Comparing the WSSV transcriptome data of shrimp to that obtained in the freshwater prawn *M. rosenbergii* against WSSV, a different gene expression pattern is observed. These results suggest that the level of transcriptomic response does not depend on the type of pathogen but on the host species, either because a host may be more susceptible or more tolerant/resistant to a pathogen.

A feature observed between *M. rosenbergii* and *L. vannamei* against WSSV infection is a decrease in expression of antiviral proteins, which could indicate that the virus is able to inhibit these proteins, or else its infection may affect the signaling for their synthesis. Nonetheless, the functional annotation of this virus has been limited by its unique sequences covering almost 90% of its genome (Rodriguez-Anaya et al., 2016). Therefore, the transcriptomic analysis of the host–pathogen interaction is the most viable alternative to determine the degree to which a crustacean is affected by WSSV. Recent data showed the mechanisms attributed to the resistance ability against WSSV by certain genotypes of penaeid shrimp (mainly *L. vannamei*). Zhan et al. (2019) showed that hemocyanin possessed antiviral functions, one of which (LvHcL48) decreased the transcription of the WSSV genes wsv069 and wsv421 and interacted with the WSSV VP28 envelope protein. Further, Zhu et al. (2019) reported that the transcription protein (STAT) signal transducer and activator is a mediator of the antiviral response of *L. vannamei* against WSSV. This protein controls the genetic modules of regulation of the viral process, the JAKSTAT cascade and the regulation of the pathways of the immune effector process. The functionality of each of the DEGs that are the product of a transcriptomic analysis appears to the observation of the pathways affected by the pathogens. Subsequent analyses will determine the possible functionality of these genes in the host–pathogen interaction. It is also possible to analyze the transcriptome of the virus, bacteria or fungus that manifests itself in crustaceans, but as in the case of WSSV, if the sequences or the functional annotation of their genes is not available, it will make the analysis of this interaction difficult.

An amazing case of *L. vannamei* susceptibility to WSSV was presented by Santos et al. (2018b), who reported 14,124 genes related to the WSSV infection. The most frequent genes in the WSSV-exposed shrimp were mainly those related to crustacyanin 1 and 2, lectins and hemocyanin B and C heavy chain. These can be compared to other farmed shrimp species such as *Fenneropenaeus chinensis*, whose gene expression values were 14,263 significant genes, expressing HSP90, calreticulin, integrin, and prophenoloxidase 2, cathepsin L, cathepsin B and protein disulfide isomerase, tubulin beta-1 chain, phosphoenolpyruvate carboxykinase and trypsin (Shi et al., 2018). Such values are similar to those described by Santos et al. (2018b) in a study on the functionality of the DEGs. These two shrimp species reacted similarly upon a WSSV infection since they displayed similar DEGs.

The Kuruma shrimp *Marsupenaeus japonicus* has also been used to report gene expression in response to WSSV. Here, 4,081 genes showed differential expression including 2,150 up-regulated and 1,931 down-regulated genes. Among the 4,081 DEGs, 426 were involved in immunomodulation, related to signal transduction, ubiquitous system, complement and coagulation cascade, phagocytosis, nervous and endocrine systems, apoptotic tumor-related protein, heat shock proteins, tentative components of the proPO system, proteinase inhibitors, peroxisome, effectors, recognition, and antioxidation. This highlights up-regulation of C-type lectin 1, suppressor of cytokine signaling, cathepsin C, fibrinogen-related proteins, ecdysone receptor, hemocyanin, heat stock transcription factor 1 and cytochrome p450. In contrast, down-regulated genes include catalase, prohibitin, caprin-1 and gamma-interferon-inducible lysosomal thiol reductase, having the lowest fold change (Zhong et al., 2017). The expression pattern observed in this species is relatively low compared to those of *F. chinensis*, *L. vannamei* and even *M. rosenbergii*.

A comparative study of all expression values could determine the importance of this pathogen and its general mechanism of action in crustaceans. For instance, comparing the DEGs in each species could help to identify genes that are directly involved with WSSV infection. Describing the functionality of each of the genes will help find potential solutions to reduce the impact of this pathogen. The results shown are a fundamental basis for the establishment of corrective and/or preventive measures, starting with tracing the transcriptomics of the interaction, followed by the formulation of potential mechanisms of action and proposing measures to control the culture conditions to modulate the transcription. The results are mainly shown in the response of the crustacean towards the presence of the pathogen, and not directly from the transcriptomics related to pathogenicity. This is because sequencing several viral genomes is limited by the presence of unique sequences, leaving the analysis of host–pathogen interaction as the only tool available to perform such studies in crustaceans. Nonetheless, the present information can serve as a basis for the identification of susceptible and resistant genotypes, allowing the use of genotypes in breeding programs. In addition, these data may also be used for the formulation of diets and supplementation of dietary additives (i.e. immunostimulants such as lipopolysaccharides or β-glucans) to induce a positive regulation of genes related to the defense response (lipopolysaccharide and β-1,3-glucan-binding protein [LGBP]) upon the presence of a pathogen (Wongsasak et al., 2015).

2.3.2.3 *Infectious Hypodermal and Hematopoietic Necrosis Virus (IHHNV)*

The infectious hypodermic and hematopoietic necrosis virus (IHHNV) has been one of the most lethal pathogens in penaeid shrimp, particularly to *L. stylirostris*, which showed mortalities of up to 90%. The shrimp species *L. vannamei* can become infected with the virus, but it is less susceptible and does not display mortality, although it may show defects such as the runt deformity syndrome. The main signs of infection in highly susceptible species are lethargy, muscle opacity and death (Vega-Heredia et al., 2012).

The impact of this virus has been extensively studied on a variety of susceptible crustaceans. These studies include heat shock proteins, which become activated when a stress condition takes place (oxidation, salinity, temperature, reduced food availability). Such studies have focused only on specific molecules such as the expression of small heat shock protein HSP37, which has been studied in various crustaceans. Arockiaraj et al. (2012c) mention that in the presence of the IHHNV, expression of HSP37 rises, indicating that this protein is associated with viral infection. But the stress induced by IHHNV infection is not only met by the heat shock proteins. Also, the expression of inhibitors of apoptosis has been reported, the function of which is to regulate cellular death. Arockiaraj et al. (2011) reported that the expression of such genes in *M. rosenbergii* infected with IHHNV showed a significant increase, indicating that these genes are involved in the immune response mechanism in this species. Likewise, caspase activation is required to trigger apoptosis. Some reports indicate that caspase is expressed in greater quantities upon IHHNV infection in *M. rosenbergii*, suggesting an interaction of caspase with the crustacean immune response (Arockiaraj et al., 2012b). Nonetheless, transcriptomic studies in these issues are limited, presenting only general information about the host–pathogen interaction and focusing on limited crustacean species (Shekhar & Ponniah, 2015).

Recent transcriptomics studies on *Procambarus clarkii* infected with IHHNV determined that the virus is associated with an increased expression of Ras signaling pathway genes (switch for proliferation, growth, migration and other cellular functions), and decreased expression of some genes in the phospholipase D pathway (involved in membrane transport, cytoskeletal remodeling, cell proliferation and cell survival) and trypsin (Nian et al., 2020). Therefore, it was found that these genes were highly related to apoptosis, phagocytosis and shrimp digestion. These results show that despite reports that the shrimp *L. stylirostris* is the most affected species, no evidence exists of changes in transcriptomic expression between crustacean species. Hence, other various crustacean species may show high levels of infection and mortality caused by the virus. This finding stresses the need to work towards the prevention of emerging diseases and their spread to other crustacean populations either wild or cultured, such as *L. vannamei*. The ability of some viruses to replicate within new hosts, especially RNA viruses, has long been known. Nonetheless, IHHNV is a single-stranded DNA virus, which shows a higher mutation rate than viruses with double-stranded DNA genomes (Sanjuán & Domingo-Calap, 2016). This feature may represent an advantage of the virus to spread and to infect other hosts, underlining the importance of studying the kinetics and mechanisms of virus infection and replication within hosts, either for the proposal of preventive measures or to establish corrective measures in cultures. Transcriptomic analyses may help to generate such data and to present evidence of the metabolic pathways involved in the crustacean responses against infection by this pathogen.

2.3.2.4 Nodavirus

Nodaviruses are seldom found in crustacean cultures. These agents have high selective pathogenicity to certain hosts. Nodaviruses affecting crustaceans cause the white tail disease. At present, two nodavirus species have been recorded affecting farmed crustaceans including the *Penaeus vannamei* nodavirus (PvNV) and the *M. rosenbergii* nodavirus (MrNV). These viruses cause a disease characterized by necrosis of striated muscle tissues in the tail of susceptible crustaceans, reporting mortalities up to 70%. A transcriptomic study on the host–pathogen interaction was done in *M. rosenbergii.* This revealed 5,538 genes related to the nodavirus infection, of which 2,413 were up-regulated and 3,125 were down-regulated. These genes were related to the immune response, showing the highest expression changes during infection. Defense-related genes included antiviral proteins, C-type lectin, prophenoloxidase, caspase and ADP ribosylation factors (Pasookhush et al., 2019; Tang et al., 2011). These nodaviruses are important in shrimp aquaculture since their presence threatens the development of the industry. Hence, more studies are needed on the transcriptomic profiles of hosts and their characterization upon nodavirus infection. Such studies will help unravel the action mechanisms of these viruses in order to propose potential control methods against them.

The pathogenicity of MrNV has been observed mainly in the early larval stages (nauplius) of crustaceans. It appears that pathogenicity of the virus decreases as the prawn grows older. Jariyapong et al. (2019) performed a transcriptomic analysis of hematopoietic tissues to determine its relationship with MrNV resistance. They observed 462 DEGs (281 positively regulated genes and 181 negatively regulated genes). Of these, a protein similar to crustacean hematopoietic factor, the anti-lipopolysaccharide factor (ALF) and the nucleolar cell growth regulatory protein from hemocyte hematopoiesis were positively regulated. This led to relate the resistance of *M. rosenbergii* to the virus with an increase in humoral immune factors and to the acceleration of hemocyte homeostasis. It is possible that improving nauplii culture conditions to boost the regulation of such genes would result in reducing the impact of this nodavirus. Nonetheless, more genes and pathways may be involved with nodavirus infection, potentially signaling the expression of those genes. It is possible that some of these genes interfere with nodavirus genes. Alternatively, the hematopoietic response observed in adult prawns could be the key in the response against MrNV. The latter is an example of the importance of transcriptomics in the host–pathogen interaction. It shows the cascade of genes involved in the infection process, and some of these genes could have a role in the resistance against the pathogen. These possibilities can be determined with additional transcriptomic studies which may lead to new data on the mechanisms of virus resistance in crustacean species.

2.3.2.5 Yellow Head Virus (YHV)

The YHV is one of the main pathogens in cultures of the black tiger shrimp (*Penaeus monodon*), which causes up to 100% mortality of infected shrimp within 3–5 days after the onset of clinical signs of virus infection (Cowley, 2016). This virus has been studied to understand the molecular mechanisms of infection in the black tiger shrimp. Transcriptomic analyses showed that 105 genes were up-regulated, of which 50 (47%) have an unknown function and the rest were related to the crustacean immune response. These genes include ferritin, carboxylesterase-6, arginase, zinc finger protein 416, nascent polypeptide-associated complex alpha chain, eukaryotic translation initiation factor, ribosomal proteins, hypothetical proteins and unknown genes (Pongsomboon et al., 2008).

Likewise, Prapavorarat et al. (2010) reported changes in gene expression levels during different phases of infection, such levels being lower at advanced phases. Genes associated with cellular defense and homeostasis were expressed in greater amounts at early stages of infection to fight YHV. Also, seven known genes were directly related to the presence of YHV (protease 3C, glycoprotein 64, glycoprotein 116, helicase, nucleocapsid protein, polyprotein replicase and RNA polymerase) and four genes related to immune response were up-regulated (ALF isoform 6, crustin isoform 1 [CrusI1] and hemocyte transglutaminase [SPIPm2]). These genes related to YHV infection have already been annotated (Sittidilokratna et al., 2008). The up-regulation of an antiviral gene of great interest in *P. monodon* has been observed. In fact, Ponprateep et al. (2012) used gene silencing to identify the factor function in the survival of *P. monodon*. Their results showed that ALF isoform 3 had an important role in hepatopancreatic bacterial and hemolymph regulation. Likewise, they observed higher mortalities of shrimp upon infection by *Vibrio harveyi* and WSSV by silencing of ALF isoform 6. These results showed that this transcript, and the family to which it belongs, has an important antimicrobial and antiviral activity, along with the activity of the gene CrusI1, which is known for its strong antimicrobial activity. This gene has been cloned for experimental purposes and it showed activity mainly against Gram-positive bacteria. Interestingly, the gene was positively regulated by the presence of a virus. Nonetheless, current works have reported new isoforms of crustin with antiviral properties (LvCrustinB in *L. vannamei*; Spcrus6 in *Scylla paramamosain*, both against WSSV) (Du et al., 2019; Li et al., 2019). These results suggest that the CrusI1 gene may have not only antimicrobial, but also antiviral activity. The analysis of sequence homology would be necessary to determine the antiviral properties in each gene. On the other hand, SPIPm2 has also been reported as a protein that actively participates in the immune response of *P. monodon*. In fact, this protein has high expression levels in hemocytes, gills and heart in shrimp infected with YHV. Silencing of this protein positively regulates the expression of apoptosis genes, two caspase isoforms, antimicrobial transcriptional factors and causes a reduction of hemocytes in the hemolymph (Visetnan et al., 2018), increasing the susceptibility to YHV.

Moreover, YHV activates the Imd signaling pathway in *P. monodon* by the presence of viruses or Gram-negative bacteria. The transcription factor NF-κB (Relish) belongs to the Imd pathway and is an important regulator in the synthesis of antimicrobial peptides. Gene expression assays of some antimicrobial peptides as penaeidin isoforms 3 and 5, ALF isoform 3 and CrusI1, showed that the latter genes were regulated by NF-κB (Relish) (Visetnan et al., 2015). This finding could explain that isoform 1 of crustin was up-regulated in the virus infection due to the expression of antimicrobial genes of the Imd signaling pathway. It is possible that the Imd pathway was activated by the YHV infection and expressed both antimicrobial and antiviral peptides, or that some structural component of the virus initiated the antimicrobial signaling of CrusI1. Nonetheless, the antiviral potential of CrusI1 is not ruled out.

2.3.2.6 Decapod Iridescent Virus 1

Novel emerging diseases such as that produced by the decapod iridescent virus (DIV1) has been the focus of recent research. This virus has been reported in shrimp and crab cultures, showing signs of infection such as anorexia and lethargy. The virus affects the lymphoid organ and myoepithelial cells in *L. vannamei*. An additional sign of disease reported by Qiu et al. (2019) is the "white head" in the freshwater prawn *M. rosenbergii*. Nonetheless, this pathogen has not been reported in wild populations of crustaceans, but reports of these infections have been done from cultures of the freshwater crayfish *Cherax quadricarinatus* in China (Qiu et al., 2021; Xu et al., 2016). One of the most important features of DIV1 infection is its relation with the expression of a triosephosphate isomerase-like gene in *L. vannamei*. A comparative analysis of the crustacean transcriptome was carried out by Liao et al. (2020a), which resulted in 1,112 DEGs, 889 up-regulated and 223 down-regulated genes, upon DIV1 infection. The metabolic pathways affected were those involved in the metabolism of fructose and mannose, inositol carbon and phosphate, and triosephosphate isomerase (TPI). Of the latter pathway, a gene silencing assay was done in *L. vannamei*: LvTPI-like, LvTPI-B-like and LvTPI-B-like1. These assays resulted in a higher mortality of shrimp infected with DIV1 but virus replication also was affected.

Other studies dealing with DIV1 infection are those of Liao et al. (2020b) and Yang et al. (2020), who analyzed the transcriptomes of *Fenneropenaeus merguiensis* and *Cherax quadricarinatus*, respectively, and observed 1,003 DEGs

(929 up-regulated and 74 down-regulated) and 1,785 DEGs (1,070 up-regulated and 715 down-regulated), respectively. Some of the DEGs included genes such as caspase, type C lectin, Wnt5, integrin, fibrinogen and LGBP, which were related to lysosome, phagosome, MAPK signaling pathway, Wnt signaling pathway and Toll-like receptor signaling pathway. It is known that many viral pathogens directly affect signaling pathways in crustaceans, since their recognition patterns allow the detection of some extracellular components characteristic of pathogens, thus initiating gene transcription in order to fight infection. Transcriptomic analyses show specifically which host genes from a metabolic pathway will respond against a pathogen. Nonetheless, genes specific to a pathogen may be unknown in several types of well-studied pathogens but also in novel emerging pathogens, but still, they will reveal the transcriptional changes.

2.3.3 Other Pathogens

2.3.3.1 *Enterocytozoon hepatopenaei*

Many pathogens have a negative influence in farmed shrimp production because they are able to cause massive mortalities. Nonetheless, some pathogens will not directly cause production losses due to mortality, but they cause chronic infections or diseases which result in losses due to hindered growth or impaired feeding. One such pathogen is the microsporidium *Enterocytozoon hepatopenaei*, which has been reported to cause severe growth impairment in *P. monodon* and *L. vannamei* farmed in Asian countries, and until recently (2017) it has been reported in Venezuela, South America (Rajendran et al., 2016; Tang et al., 2017). Although clinical signs of infection are limited to growth, an additional sign reported recently is the appearance of white feces in shrimp infected with the fungus-related organism. The occurrence of the pathogen and its association with the white feces was previously reported by Tangprasittipap et al. (2013). Nonetheless, they concluded that the appearance of white feces must be the result of another underlying cause and not only by the presence of *E. hepatopenaei*.

At present, and to the best of our knowledge, no studies have analyzed the transcriptome of shrimp infected by this pathogen. A study conducted by Ning et al. (2019) used omics such as metabolomics and proteomics to determine the effect of this pathogen on *L. vannamei*. Their results showed a higher immunological activity and better detection of shrimp pathogens. Further, the microsporidium infection led to a decrease in metabolic activity and energy, inducing growth hormone disorders and preventing molting. The transcriptomic analysis of this interaction may result in a better understanding of the genetic background of *E. hepatopenaei* and its influence on the shrimp transcriptome. This would make it possible to determine the pathogenicity mechanisms of the microorganism and its relationship with the presence of white feces. Also, transcriptomics, along with the information already available, will help to determine whether the pathogen is able to down-regulate the expression of host genes in important metabolic pathways.

2.3.3.2 *Spiroplasma eriocheir*

At the beginning of the 21st century, the bacterium *Spiroplasma eriocheir* was identified as the causative agent of the tremor disease in the Chinese mitten crab *Eriocheir sinensis* (Wang et al., 2003; Wang et al., 2004). Previously, outbreaks of this pathogenic bacterium were reported with mortalities between 10% and 90% in *L. vannamei* farmed in Colombia. Sequence analyses showed that the pathogen affecting farmed shrimp in Colombia had 99% similarity in the 16S ribosomal RNA with *Spiroplasma insolitum* (Nunan et al., 2005). The latter species has been reported as an endosymbiont bacterium in insects that were resistant against pathogens (Chepkemoi et al., 2017). Due to the importance of bacteria of the genus *Spiroplasma*, various studies have been carried out to study their interaction with different hosts. Recently, Hou et al. (2020) published the transcriptome of the thoracic ganglion of *E. sinensis* in interaction with *S. eriocheir*. Results produced 8,049 DEGs (927 up-regulated and 7,122 down-regulated), the proPO system being one of the pathways that were most related with *S. eriocheir* infection. In addition, two types of type C lectin were associated as receptors and activators of the proPO system. On the other hand, Ren et al. (2020) obtained the transcriptome of *L. vannamei* when interacted with *S. eriocheir*. Results showed 1,168 DEGs (792 up-regulated and 376 down-regulated) involved in the immune response and stress, the cell cycle and the organelle organization. The appearance of *Spiroplasma* in crustacean cultures has been regarded as a potentially emerging disease, as quite heterogeneous interactions between the different crustacean species and the pathogens have been reported. Nonetheless, the bacterium-influenced gene expression, especially in *E. sinensis*, inducing a negative gene regulation in almost 90% of all differentially expressed genes. It is possible that this bacterium induces expression inhibitors or disrupts the gene regulation of crustaceans, giving rise to the tremor disease.

The work of Ren et al. (2020) analyzed long-chain non-coding RNA (lncRNA), observing a correlation of 22 lncRNAs and 195 unigenes including those of proPO and LGBP. The lncRNAs make up a vast array of unexplored information found in the non-coding regions of eukaryotic cells, and they have been linked to imprinting of genomic loci, chromosome conformation, and allosteric regulation of enzyme activity (Kung et al., 2013; Quinn and Chang, 2016). This finding expands the range of unexplored and unknown information that can be gathered and analyzed with transcriptomics, much alike as miRNAs participate in genetic and cellular regulation processes and potentially play an important role in gene regulation.

2.4 Transcriptomics and MicroRNAs in Crustaceans

Regulation of gene expression is an important mechanism that can control several pathways; therefore, it is a process that needs to be controlled. Micro-RNAs (miRNAs) are small non-coding RNAs, with 18–28 nt in length, that can regulate gene expression post-transcriptionally by coupling with their complementary sequences located at the 3′ untranslated mRNA regions, ending translation by inhibiting mRNAs (Tanase et al., 2012; Wei & Wong, 2013). For the deep study of the transcriptome, several methodologies have been implemented to identify miRNAs, which have been studied for its correlation with host–pathogen interactions and immune response mechanisms in crustaceans. However, this technology has focused mainly on viruses, WSSV being the most studied. The appeal of miRNA research has focused over the last few decades on the discovery and description of pathogenicity mechanisms. Despite this, limited information exists on miRNA's implementation.

Research done by Huang et al. (2012) revealed 31 miRNAs differentially expressed in *M. japonicus* infected with WSSV. These miRNAs were involved in virus-host interactions related to signaling the immune response. In a later work, they demonstrated that one of the differentially expressed miRNAs (miR-7) played a more important role in the response against WSSV. By blocking this miRNA, the amount of viral genome copies in the shrimp increased 10 times (Huang & Zhang, 2012). Reports have been done on miRNAs expressed in *Vibrio alginolyticus* infections in *M. rosenbergii*. These gene expression studies showed 55 differentially expressed miRNAs found in this shrimp species upon *Vibrio* infection. Of these, 8 genes were up-regulated and 30 down-regulated in the host. Such genes were related to apoptosis, cell differentiation and proliferation, energy metabolism and cancer (Zhu et al., 2016).

The miRNAs have a very important role in the crustacean transcriptome. Although they have been studied mainly in regard to signaling and gene regulation, they have an essential role in the cell as they are required to establish intercellular communications. Also, miRNAs generate a lot of information that has been scarcely explored and applied in the overall study of crustaceans and their interaction with pathogens. Studies dealing with diseases caused by viruses, bacteria and fungi that affect crustacean cultures have produced information that can be related to the transcriptome. Nonetheless, of all the work done in the subject, only a small part is related to the noncoding regions of the transcripts, which are the main source of miRNA's information. Hence, a large amount of information on miRNAs remains necessary to generate in order to determine their effect on regulation of the host–pathogen interaction transcriptome in farmed crustaceans. Very scarce information exists on the main cultured crustaceans such as *L. vannamei, P. monodon, Procambarus clarkii* and *M. rosenbergii* and their respective host–pathogen interactions in terms of miRNAs. This huge gap in transcriptomic information is another opportunity for research, and it will allow gathering of knowledge in order to reduce the impact of pathogens on aquaculture operations.

2.5 Perspectives

The determination of specific activity of crustaceans against pathogens would serve as a basis to create biomarkers that will allow: (1) identifying of crustacean genotypes and their relationship with the presence of viral or bacterial pathogens, in order to select and to make a genetic classification of the families that may be susceptible, tolerant or resistant against a pathogen; (2) isolating and characterizing pathogenic strains of infectious agents for the determination of the pathogen genotypes that may be the greatest threats on farmed crustacean health; (3) generating inhibitors for specific transcript sequences of highly pathogenic bacteria and viruses.

The results of transcriptomic analyses directed towards host–pathogen interactions can be used to determine the characteristics of each pathogen in order to propose effective control methods. Vast information exists on some specific host–pathogen interactions, but this is not directly related to the identification of genotypes. This makes it necessary to conduct studies aimed at identifying crustacean genotypes that may be resistant to pathogens, as an initial step towards the development of selective breeding programs using genetic lines with low susceptibility and enhanced transcriptomic profiles and defense pathways. Also, the functional description of transcriptomic responses will need to be developed in order to fully understand host–pathogen interaction mechanisms.

Much of the research efforts on farmed shrimp species such as *L. vannamei* has been done on defense response and differential gene expression upon pathogen infections. Nonetheless, more power needs to be directed towards transcriptomic research in other farmed crustaceans in order to unveil the mechanisms to improve the crustacean defense response upon viral or bacterial infections. Also, it is important to make transcriptomic studies on the interaction between novel emerging pathogens and the various crustacean hosts. Further, additional transcriptomic studies are needed on the defense response of crustaceans in order to determine which genes or physiological pathways can be useful in the selection of disease-resistant (or disease-tolerant) families in breeding selection programs.

Next-generation sequencing (NGS) has become a very useful tool to generate novel knowledge about genomic behavior related to various aspects of the defense response and host genetic resistance or tolerance to pathogens. These tools open the door to investigate other RNA transcription and regulation systems such as miRNAs in these animals. Transcriptomics are also useful for the description of utility, structure and evolution of specific genes, in order to understand the mechanisms in which each crustacean reacts to different pathogens. Some shrimp families and even different species are selectively resistant to some high-impact pathogens. This fact proposes the possibility that certain genes or genetic features could be selected in order to produce animals that can withstand a number of infectious diseases and even be able to stand against stressful conditions. The use of NGS will foster the knowledge of transcriptomes in different species and will allow replication of some mechanisms of action against infectious diseases.

References

Aoki T, Wang HC, Unajak S, Santos MD, Kondo H, Hirono I (2011) Microarray analyses of shrimp immune responses. *Marine Biotechnology* 13(4): 629–638. doi:10.1007/s10126-010-9291-1.

Arockiaraj J., Avin FA, Vanaraja P, Easwvaran S, Singh A, Othman RY, Bhassu S (2012a). Immune role of MrNFκBI-α, an IκB family member characterized in prawn *M. rosenbergii*. *Fish & Shellfish Immunology* 33(3):619–625. doi:10.1016/j.fsi.2012.06.015.

Arockiaraj J, Easwvaran S, Vanaraja P, Singh A, Othman RY, Bhassu S (2012b) Effect of infectious hypodermal and haematopoietic necrosis virus (IHHNV) infection on caspase 3c expression and activity in freshwater prawn *Macrobrachium rosenbergii*. *Fish and Shellfish Immunology* 32(1): 161–169. doi:10.1016/j.fsi.2011.11.006.

Arockiaraj J, Vanaraja P, Easwvaran S, Singh A, Othman RY, Bhassu S (2012c) Gene expression and functional studies of small heat shock protein 37 (MrHSP37) from *Macrobrachium rosenbergii* challenged with infectious hypodermal and hematopoietic necrosis virus (IHHNV). *Molecular Biology Reports* 39(6): 6671–6682. doi:10.1007/s11033-012-1473-7.

Arockiaraj J, Gnanam AJ, Muthukrishnan D, Gudimella R, Milton J, Singh A, Muthupandian S, Kasi M, Bhassu S (2013a) Crustin, a WAP domain containing antimicrobial peptide from freshwater prawn *Macrobrachium rosenbergii*: Immune characterization. *Fish and Shellfish Immunology* 34(1): 109–118. doi:10.1016/j.fsi.2012.10.009.

Arockiaraj J, Gnanam AJ, Palanisamy R, Kumaresan V, Bhatt P, Thirumalai MK, Roy A, Pasupuleti M, Kasi M, Sathyamoorthi A, Arasu A (2013b) A prawn transglutaminase: Molecular characterization and biochemical properties. *Biochimie* 95(12): 2354–2364. doi:10.1016/j.biochi.2013.08.029.

Arockiaraj J, Vanaraja P, Easwvaran S, Singh A, Othman RY, Bhassu S (2011) Bioinformatic characterization and gene expression pattern of apoptosis inhibitor from *Macrobrachium rosenbergii* challenged with infectious hypodermal and hematopoietic necrosis virus. *Fish and Shellfish Immunology* 31(6): 1259–1267. doi:10.1016/j.fsi.2011.09.008.

Bi J, Ning M, Xie X, Fan W, Huang Y, Gu W, Wang W, Wang L, Meng Q (2020) A typical C-type lectin, perlucin-like protein, is involved in the innate immune defense of whiteleg shrimp *Litopenaeus vannamei*. *Fish and Shellfish Immunology* 103: 293–301. doi:10.1016/j.fsi.2020.05.046.

Brubaker SW, Bonham KS, Zanoni I, Kagan JC (2015) Innate immune pattern recognition: A cell biological perspective. *Annual Review of Immunology* 33: 257–290. doi:10.1146/annurev-immunol-032414-112240.

Campos-García JC, Rodríguez-Ramírez R, Avila-Villa LA, Gómez-Aldama O, Arias-Martínez J, Gollas-Galván T (2019) Fractal dimension of hepatopancreas of white shrimp *Litopenaeus vannamei* infected with *Hepatobacter penaei* bacteria (NHPB). *Aquaculture International* 28(2020): 661–673. doi:10.1007/s10499-019-00487-y.

Chai CY, Yoon J, Lee YS, Kim YB, Choi TJ (2013) Analysis of the complete nucleotide sequence of a white spot syndrome virus isolated from pacific white shrimp. *Journal of Microbiology* 51(5): 695–699. doi:10.1007/s12275-013-3171-0.

Chaurasia MK, Nizam F, Ravichandran G, Arasu MV, Al-Dhabi NA, Arshad A, Elumalai P, Arockiaraj J (2016) Molecular importance of prawn large heat shock proteins 60, 70 and 90. *Fish and Shellfish Immunology* 48: 228–238. doi:10.1016/j.fsi.2015.11.034.

Chen Xiaohan, Zeng D, Chen Xiuli, Xie D, Zhao Y, Yang C, Li Y, Ma N, Li M, Yang Q, et al. (2013) Transcriptome analysis of *Litopenaeus vannamei* in response to white spot syndrome virus infection. *PLoS ONE* 8(8). doi:10.1371/journal.pone.0073218.

Chen YH, Yuan FH, Bi HT, Zhang ZZ, Yue HT, Yuan K, Chen YG, Wen SP, He JG (2016) Transcriptome analysis of the unfolded protein response in hemocytes of *Litopenaeus vannamei*. *Fish and Shellfish Immunology* 54: 153–163. doi:10.1016/j.fsi.2015.10.027.

Chepkemoi ST, Mararo E, Butungi H, Paredes J, Masiga D, Sinkins SP, Herren JK (2017) Identification of *Spiroplasma insolitum* symbionts in *Anopheles gambiae*. *Wellcome Open Research* 2(90). doi:10.12688/wellcomeopenres.12468.1.

Chu D, Wei L (2019) Nonsynonymous, synonymous and nonsense mutations in human cancer-related genes undergo stronger purifying selections than expectation. *BMC Cancer* 19(1): 1–12. doi:10.1186/s12885-019-5572-x.

Cowley J A (2016) Nidoviruses of fish and crustaceans. In: *Aquaculture Virology*, pp. 443–472. Academic Press.

Du Z-Q, Wang Y, Ma H-Y., Shen X-L, Wang K, Du J, Yu X-D, Fang W-H, Li X-C (2019) A new crustin homologue (SpCrus6) involved in the antimicrobial and antiviral innate immunity in mud crab, *Scylla paramamosain*. *Fish and Shellfish Immunology* 84: 733–743. doi:10.1016/j.fsi.2018.10.072.

FAO (2018) *Estado Mundial de la Pesca y la Acuicultura*. Organización de las Naciones Unidas para la Alimentación y la *Agricultura*: 19–60.

Figueroa-Pizano MD, Peregrino-Uriarte AB, Yepiz-Plascencia G, Martínez-Porchas M, Gollas-Galván T, Martínez-Córdova LR (2014) Gene expression responses of white shrimp (*Litopenaeus vannamei*) infected with necrotizing hepatopancreatitis bacterium. *Aquaculture* 420–421: 165–170. doi:10.1016/j.aquaculture.2013.10.042.

Flegel TW (2012) Historic emergence, impact and current status of shrimp pathogens in Asia. *Journal of Invertebrate Pathology* 110(2): 166–173. doi:10.1016/j.jip.2012.03.004.

Gatto M, Iaccarino L, Ghirardello A, Bassi N, Pontisso P, Punzi L, Shoenfeld Y, Doria A (2013) Serpins, immunity and autoimmunity: Old molecules, new gunctions. *Clinical Reviews in Allergy & Immunology* 45(2): 267–280. doi:10.1007/s12016-013-8353-3.

Gulati A, Kumar R, Mukhopadhaya A (2019) Differential recognition of *Vibrio parahaemolyticus* OmpU by Toll-like receptors in monocytes and macrophages for the induction of proinflammatory responses. *Infection and Immunity* 87(5). doi:10.1128/IAI.00809-18.

Güiza J, Barría I, Sáez JC and Vega JL (2018) Innexins: Expression, regulation, and functions. *Frontiers in Physiology* 9: 1414. doi.org/10.3389/fphys.2018.01414.

Habib YJ, Zhang Z (2020) The involvement of crustaceans toll-like receptors in pathogen recognition. *Fish and Shellfish Immunology* 102: 169–176. doi:10.1016/j.fsi.2020.04.035.

Hijikata A, Kitamura H, Kimura Y, Yokoyama R, Aiba Y, Bao Y, Fujita S, Hase K, Hori S, Ishii Y, et al. (2007) Construction of an open-access database that integrates cross-reference information from the transcriptome and proteome of immune cells. *Bioinformatics* 23(21): 2934–2941. doi:10.1093/bioinformatics/btm430.

Horgan RP, Kenny LC (2011) "Omic" technologies: Genomics, transcriptomics, proteomics and metabolomics. *The Obstetrician & Gynaecologist* 13(3): 189–195. doi:10.1576/toag.13.3.189.27672.

Hou L, Ma Y, Cao X, Gu W, Cheng Y, Wu X, Wang W, Meng Q (2020) Transcriptome profiling of the *Eriocheir sinensis* thoracic ganglion under the *Spiroplasma eriocheiris* challenge. *Aquaculture* 524(December 2019): 735257. doi:10.1016/j.aquaculture.2020.735257.

Huang T, Xu D, Zhang X (2012) Characterization of host microRNAs that respond to DNA virus infection in a crustacean. *BMC Genomics* 13(1). doi:10.1186/1471-2164-13-159.

Huang T, Zhang X (2012) Functional analysis of a crustacean MicroRNA in host-virus interactions. *Journal of Virology* 86(23): 12997–13004. doi:10.1128/jvi.01702-12.

Jariyapong P, Pudgerd A, Cheloh N, Hirono I, Kondo H, Vanichviriyakit R, Weerachatyanukul W, Chotwiwatthanakun C (2019) Hematopoietic tissue of *Macrobrachium rosenbergii* plays dual roles as a source of hemocyte hematopoiesis and as a defensive mechanism against *Macrobrachium rosenbergii* nodavirus infection. *Fish and Shellfish Immunology* 86: 756–763. doi:10.1016/j.fsi.2018.12.021.

Junprung W, Supungul P, Tassanakajon A (2017) HSP70 and HSP90 are involved in shrimp *Penaeus vannamei* tolerance to AHPND-causing strain of *Vibrio parahaemolyticus* after non-lethal heat shock. *Fish and Shellfish Immunology* 60: 237–246. doi:10.1016/j.fsi.2016.11.049.

Kang R, Zeh HJ, Lotze MT and Tang D (2011) The Beclin 1 network regulates autophagy and apoptosis. *Cell Death and Differentiation* 18(4): 571–580. doi.org/10.1038/cdd.2010.191.

Kung JTY, Colognori D, Lee JT (2013) Long noncoding RNAs: Past, present, and future. *Genetics* 193(3): 651–669. doi:10.1534/genetics.112.146704.

Li C, Weng S, Chen Y, Yu X, Lü L, Zhang H, He J, Xu X (2012) Analysis of *Litopenaeus vannamei* transcriptome using the next-generation DNA sequencing technique. *PLoS ONE* 7(10). doi:10.1371/journal.pone.0047442.

Li E, Wang X, Chen K, Xu C, Qin JG, Chen L (2017) Physiological change and nutritional requirement of Pacific white shrimp *Litopenaeus vannamei* at low salinity. *Reviews in Aquaculture* 9(1): 57–75. doi:10.1111/raq.12104.

Li M, Ma C, Zhu P, Yang Y, Lei A, Chen X, Liang W, Chen M, Xiong J, Li C (2019) A new crustin is involved in the innate immune response of shrimp *Litopenaeus vannamei*. *Fish and Shellfish Immunology* 94: 398–406. doi:10.1016/j.fsi.2019.09.028.

Li S, Zhang X, Sun Z, Li F, Xiang J (2013) Transcriptome analysis on Chinese shrimp *Fenneropenaeus chinensis* during WSSV acute infection. *PLoS ONE* 8(3). doi:10.1371/journal.pone.0058627.

Liao X, Wang C, Wang B, Qin H, Hu S, Wang P, Sun C, Zhang S (2020a) Comparative transcriptome analysis of *Litopenaeus vannamei* reveals that triosephosphate isomerase-like genes play an important role during Decapod Iridescent Virus 1 *Infection*. *Frontiers in Immunology* 11: 1–17. doi:10.3389/fimmu.2020.01904.

Liao X Zheng, Wang C Gui, Wang B, Qin H Peng, Hu S Kang, Zhao J Chen, He Z Hao, Zhong Y Qi, Sun C Bo, Zhang S (2020b) Research into the hemocyte immune response of *Fenneropenaeus merguiensis* under decapod iridescent virus 1 (DIV1) challenge using transcriptome analysis. *Fish and Shellfish Immunology* 104: 8–17. doi:10.1016/j.fsi.2020.05.053.

Lightner DV, Redman RM, Pantoja CR, Tang KFJ, Noble BL, Schofield P, Mohney LL, Nunan LM, Navarro SA (2012) Historic emergence, impact and current status of shrimp pathogens in the Americas. *Journal of Invertebrate Pathology* 110(2): 174–183. doi:10.1016/j.jip.2012.03.006.

Ma K-W, Ma W (2016) YopJ family effectors promote bacterial infection through a unique acetyltransferase activity. *Microbiology and Molecular Biology Reviews* 80(4): 1011–1027. doi:10.1128/MMBR.00032-16.

Maralit BA, Jaree P, Boonchuen P, Tassanakajon A, Somboonwiwat K (2018) Differentially expressed genes in hemocytes of *Litopenaeus vannamei* challenged with *Vibrio parahaemolyticus* AHPND (VPAHPND) and VPAHPND toxin. *Fish & Shellfish Immunology* 81: 284–296. doi:10.1016/j.fsi.2018.06.054.

Nian YY, Chen BK, Wang JJ, Zhong WT, Fang Y, Li Z, Zhang QS, Yan DC (2020) Transcriptome analysis of *Procambarus clarkii* infected with infectious hypodermal and haematopoietic necrosis virus. *Fish and Shellfish Immunology* 98: 766–772. doi:10.1016/j.fsi.2019.11.027.

Ning M, Wei P, Shen H, Wan X, Jin M, Li X, Shi H, Qiao Y, Jiang G, Gu W, et al. (2019) Proteomic and metabolomic responses in hepatopancreas of whiteleg shrimp *Litopenaeus vannamei* infected by microsporidian *Enterocytozoon hepatopenaei*. *Fish and Shellfish Immunology* 87: 534–545. doi:10.1016/j.fsi.2019.01.051.

Nunan LM, Lightner DV, Oduori MA, Gasparich GE (2005) *Spiroplasma penaei* sp. nov., associated with mortalities in *Penaeus vannamei*, Pacific white shrimp. *International Journal of Systematic and Evolutionary Microbiology* 55(6): 2317–2322. doi:10.1099/ijs.0.63555-0.

Nunan LM, Pantoja CR, Gomez-Jimenez S, Lightner DV (2012) "Candidatus Hepatobacter penaei," an intracellular pathogenic enteric bacterium in the hepatopancreas of the marine shrimp *Penaeus vannamei* (Crustacea: Decapoda). *Applied and Environmental Microbiology* 79(4): 1407–1409. doi:10.1128/aem. 02425-12.

Page RB, Stromberg AJ (2011) Linear methods for analysis and quality control of relative expression ratios from quantitative real-time polymerase chain reaction experiments. *Scientific World Journal* 11: 1383–1393. doi:10.1100/tsw.2011.124.

Pasookhush P, Hindmarch C, Sithigorngul P, Longyant S, Bendena WG, Chaivisuthangkura P (2019) Transcriptomic analysis of *Macrobrachium rosenbergii* (giant fresh water prawn) post-larvae in response to *M. rosenbergii* nodavirus

(MrNV) infection: De novo assembly and functional annotation. *BMC Genomics* 20(1): 762. doi:10.1186/s12864-019-6102-6.

Pfaffl, MW (2001). A new mathematical model for relative quantification in real-time RT–PCR. *Nucleic Acids Research* 29(9): article e45.

Picard B, Lebret B, Cassar-Malek I, Liaubet L, Berri C, Le Bihan-Duval E, Hocquette JF, Renand G (2015) Recent advances in omic technologies for meat quality management. *Meat Science* 109: 18–26. doi:10.1016/j.meatsci.2015.05.003.

Pongsomboon S, Tang S, Boonda S, Aoki T, Hirono I, Yasuike M, Tassanakajon A (2008) Differentially expressed genes in *Penaeus monodon* hemocytes following infection with yellow head virus. *Journal of Biochemistry and Molecular Biology* 41(9): 670–677. doi:10.5483/bmbrep.2008.41.9.670.

Ponprateep S, Tharntada S, Somboonwiwat K, Tassanakajon A (2012) Gene silencing reveals a crucial role for anti-lipopolysaccharide factors from *Penaeus monodon* in the protection against microbial infections. *Fish and Shellfish Immunology* 32(1): 26–34. doi:10.1016/j.fsi.2011.10.010.

Prapavorarat A, Pongsomboon S, Tassanakajon A (2010) Identification of genes expressed in response to yellow head virus infection in the black tiger shrimp, *Penaeus monodon*, by suppression subtractive hybridization. *Developmental and Comparative Immunology* 34(6): 611–617. doi:10.1016/j.dci.2010.01.002.

Qin Z, Babu VS, Wan Q, Zhou M, Liang R, Muhammad A, Zhao L, Li J, Lan J, Lin, L (2018) Transcriptome analysis of Pacific white shrimp (*Litopenaeus vannamei*) challenged by Vibrio parahaemolyticus reveals unique immune-related genes. *Fish & Shellfish Immunology* 77: 164–174. doi:10.1016/j.fsi.2018.03.030.

Qiu L, Chen X, Gao W, Li C, Guo XM, Zhang QL, Yang B, Huang J (2021) Molecular epidemiology and histopathological study of a natural infection with decapod iridescent virus 1 in farmed white leg shrimp, *Penaeus vannamei*. *Aquaculture* 533(106): 736105. doi:10.1016/j.aquaculture.2020.736105.

Qiu L, Chen X, Zhao RH, Li C, Gao W, Zhang QL, Huang J (2019) Description of a natural infection with decapod iridescent virus 1 in farmed giant freshwater prawn, *Macrobrachium rosenbergii*. *Viruses* 11(4): 1–14. doi:10.3390/v11040354.

Quinn JJ, Chang HY (2016) Unique features of long non-coding RNA biogenesis and function. *Nature Reviews Genetics* 17(1): 47–62. doi:10.1038/nrg.2015.10.

Rajendran KV, Shivam S, Ezhil Praveena P, Rajan JS, Kumar TS, Avunje S, Jagadeesan V, Prasad Babu SVANV, Pande A, Krishnan AN, et al. (2016) Emergence of *Enterocytozoon hepatopenaei* (EHP) in farmed *Penaeus* (*Litopenaeus*) *vannamei* in India. *Aquaculture* 454(2016): 272–280. doi:10.1016/j.aquaculture.2015.12.034.

Rao R, Bhassu S, Bing RZY, Alinejad T, Hassan SS, Wang J (2016) A transcriptome study on *Macrobrachium rosenbergii* hepatopancreas experimentally challenged with white spot syndrome virus (WSSV). *Journal of Invertebrate Pathology* 136: 10–22. doi:10.1016/j.jip.2016.01.002.

Rao R, Bing Zhu Y, Alinejad T, Tiruvayipati S, Lin Thong K, Wang J, Bhassu S (2015) RNA-seq analysis of *Macrobrachium rosenbergii* hepatopancreas in response to *Vibrio parahaemolyticus* infection. *Gut Pathogens* 7(1): 1–16. doi:10.1186/s13099-015-0052-6.

Ravichandran G, Kumaresan V, Mahesh A, Dhayalan A, Arshad A, Arasu MV, Al-Dhabi NA, Pasupuleti M, Arockiaraj J (2018) Bactericidal and fungistatic activity of peptide derived from GH18 domain of prawn chitinase 3 and its immunological functions during biological stress. *International Journal of Biological Macromolecules* 106: 1014–1022. doi:10.1016/j.ijbiomac.2017.08.098.

Ren Q, Zhang Z, Li X, Jie-Du Hui KM, Zhang CY, Wang W (2012) Three different anti-lipopolysaccharide factors identified from giant freshwater prawn, *Macrobrachium rosenbergii*. *Fish and Shellfish Immunology* 33(4): 766–774. doi:10.1016/j.fsi.2012.06.032.

Ren Y, Li J, Guo L, Liu JN, Wan H, Meng Q, Wang H, Wang Z, Lv L, Dong X, et al. (2020) Full-length transcriptome and long non-coding RNA profiling of whiteleg shrimp *Penaeus vannamei* hemocytes in response to *Spiroplasma eriocheiris* infection. *Fish and Shellfish Immunology* 106: 876–886. doi:10.1016/j.fsi.2020.06.057.

Rodriguez-Anaya LZ, Casillas-Hernandez R, Lares-Villa L, Gonzalez-Galaviz JR (2018) Next-generation sequencing technologies and the improvement of aquaculture sustainability of Pacific white shrimp (*Litopenaeus vannamei*). *Current Science* 115(2): 202–203.

Rodriguez-Anaya LZ, Gonzalez-Galaviz JR, Casillas-Hernandez R, Lares-Villa F, Estrada K, Ibarra-Gamez JC, Sanchez-Flores A (2016) Draft genome sequence of white spot syndrome virus isolated from cultured *Litopenaeus vannamei* in Mexico. *Genome Announcements* 4(2): 4–5. doi:10.1128/genomeA.01674-15.

Sanjuán R, Domingo-Calap P (2016) Mechanisms of viral mutation. *Cellular and Molecular Life Sciences* 73(23): 4433–4448. doi:10.1007/s00018-016-2299-6.

Santos CA, Andrade SCS, Freitas PD (2018a) Identification of SNPs potentially related to immune responses and growth performance in *Litopenaeus vannamei* by RNA-seq analyses. *Peer Journal* 2018a(7): 1–19. doi:10.7717/peerj.5154.

Santos CA, Andrade SCS, Teixeira AK, Farias F, Kurkjian K, Guerrelhas AC, Rocha JL, Galetti PM, Freitas PD (2018b) *Litopenaeus vannamei* transcriptome profile of populations evaluated for growth performance and exposed to white spot syndrome virus (WSSV). *Frontiers in Genetics* 9: 1–6. doi:10.3389/fgene.2018.00120.

Shekhar MS, Ponniah AG (2015) Recent insights into host–pathogen interaction in white spot syndrome virus infected penaeid shrimp. *Journal of Fish Diseases* 38(7): 599–612. doi:10.1111/jfd.12279.

Shi X, Meng X, Kong J, Luan S, Luo K, Cao B, Lu X, Li X, Chen B, Cao J (2018) Transcriptome analysis of "Huanghai No. 2" *Fenneropenaeus chinensis* response to WSSV using RNA-seq. *Fish and Shellfish Immunology* 75: 132–138. doi:10.1016/j.fsi.2018.01.045.

Sittidilokratna N, Dangtip S, Cowley JA, Walker PJ (2008) RNA transcription analysis and completion of the genome sequence of yellow head nidovirus. *Virus Research* 136(1–2): 157–165. doi:10.1016/j.virusres.2008.05.008.

Sookruksawong S, Sun F, Liu Z, Tassanakajon A (2013) RNA-Seq analysis reveals genes associated with resistance to Taura syndrome virus (TSV) in the Pacific white shrimp *Litopenaeus vannamei*. *Developmental and Comparative Immunology* 41(4): 523–533. doi:10.1016/j.dci.2013.07.020.

Suravajhala P, Kogelman LJA, Kadarmideen HN (2016) Multi-omic data integration and analysis using systems genomics approaches: Methods and applications in animal production, health and welfare. *Genetics Selection Evolution* 48(1): 1–14. doi:10.1186/s12711-016-0217-x.

Tanase CP, Ogrezeanu I, Badiu C (2012) MicroRNAs. *Molecular Pathology of Pituitary Adenomas* 91–96. doi:10.1016/b978-0-12-415830-6.00008-1.

Tang KFJ, Aranguren LF, Piamsomboon P, Han JE, Maskaykina IY, Schmidt MM (2017) Detection of the microsporidian *Enterocytozoon hepatopenaei* (EHP) and Taura syndrome virus in *Penaeus vannamei* cultured in Venezuela. *Aquaculture* 480: 17–21. doi:10.1016/j.aquaculture.2017.07.043.

Tang KFJ, Pantoja CR, Redman RM, Navarro SA, Lightner DV (2011) Ultrastructural and sequence characterization of *Penaeus vannamei* nodavirus (PvNV) from Belize. *Diseases of Aquatic Organisms* 94(3): 179–187. doi:10.3354/dao02335.

Tangprasittipap A, Srisala J, Chouwdee S, Somboon M, Chuchird N, Limsuwan C, Srisuvan T, Flegel TW, Sritunyalucksana K (2013) The microsporidian *Enterocytozoon hepatopenaei* is not the cause of white feces syndrome in whiteleg shrimp *Penaeus (Litopenaeus) vannamei*. *BMC Veterinary Research* 9. doi:10.1186/1746-6148-9-139.

Trang TT, Hung NH, Ninh NH, Knibb W, Nguyen NH (2019) Genetic variation in disease resistance against white spot syndrome virus (WSSV) in *Litopenaeus vannamei*. *Frontiers in Genetics* 10: 1–10. doi:10.3389/fgene.2019.00264.

Trosky JE, Mukherjee S, Burdette DL, Roberts M, McCarter L, Siegel RM, Orth K (2004) Inhibition of MAPK signaling pathways by VopA from *Vibrio parahaemolyticus*. *Journal of Biological Chemistry* 279(50): 51953–51957. doi:10.1074/jbc. M407001200.

Vega-Heredia S, Mendoza-Cano F, Sánchez-Paz A (2012) The infectious hypodermal and haematopoietic necrosis virus: A brief review of what we do and do not know. *Transboundary and Emerging Diseases* 59(2): 95–105. doi:10.1111/j.1865-1682.2011.01249.x.

Veloso A, Warr GW, Browdy CL, Chapman RW (2011) The transcriptomic response to viral infection of two strains of shrimp (*Litopenaeus vannamei*). *Developmental and Comparative Immunology* 35(3): 241–246. doi:10.1016/j.dci.2010.10.001.

Visetnan S, Supungul P, Tang S, Hirono I, Tassanakajon A, Rimphanitchayakit V (2015) YHV-responsive gene expression under the influence of Pm Relish regulation. *Fish & Shellfish Immunology* 47(1): 572–581. doi:10.1016/j.fsi.2015.09.053.

Visetnan S, Donpudsa S, Tassanakajon A, Rimphanitchayakit V (2018) Silencing of a Kazal-type serine proteinase inhibitor SPIPm2 from *Penaeus monodon* affects YHV susceptibility and hemocyte homeostasis. *Fish and Shellfish Immunology* 79: 18–27. doi:10.1016/j.fsi.2018.05.004.

Wang W, Rong L, Gu W, Du K, Chen J (2003) Study on experimental infections of *Spiroplasma* from the Chinese mitten crab in crayfish, mice and embryonated chickens. *Research in Microbiology* 154(10): 677–680. doi:10.1016/j. resmic.2003.08.004.

Wang W, Wen B, Gasparich GE, Zhu N, Rong L, Chen J, Xu Z (2004) A *Spiroplasma* associated with tremor disease in the Chinese mitten crab (*Eriocheir sinensis*). *Microbiology* 150(9): 3035–3040. doi:10.1099/mic.0.26664-0.

Wang XW, Wang JX (2013) Pattern recognition receptors acting in innate immune system of shrimp against pathogen infections. *Fish and Shellfish Immunology* 34(4): 981–989. doi:10.1016/j.fsi.2012.08.008.

Wei X, Liu X, Yang Jianmin, Fang J, Qiao H, Zhang Y, Yang Jialong (2012) Two C-type lectins from shrimp *Litopenaeus vannamei* that might be involved in immune response against bacteria and virus. *Fish and Shellfish Immunology* 32(1): 132–140. doi:10.1016/j.fsi.2011.11.001.

Wei Z, Wong GW (2013) Roles of Micro-RNAs in metabolism. *Encyclopedia of Biological Chemistry* 191–194. doi:10.1016/b978-0-12-378630-2.00078-5.

Wongsasak U, Chaijamrus S, Kumkhong S, Boonanuntanasarn S (2015) Effects of dietary supplementation with β-glucan and synbiotics on immune gene expression and immune parameters under ammonia stress in Pacific white shrimp. *Aquaculture* 436: 179–187. doi:10.1016/j.aquaculture.2014.10.028.

Wu M, Jin X, Yu A, Zhu Y, Li D, Li W, Wang Q (2014) Caspase-mediated apoptosis in crustaceans: Cloning and functional characterization of EsCaspase-3-like protein from *Eriocheir sinensis*. *Fish and Shellfish Immunology* 41(2): 625–632. doi:10.1016/j.fsi.2014.10.017.

Xu L, Wang T, Li F, Yang F (2016) Isolation and preliminary characterization of a new pathogenic iridovirus from redclaw crayfish *Cherax quadricarinatus*. *Diseases of Aquatic Organisms* 120(1): 17–26. doi:10.3354/dao03007.

Xue S, Liu Y, Zhang Y, Sun Y, Geng X, Sun J (2013) Sequencing and De novo analysis of the hemocytes transcriptome in *Litopenaeus vannamei* response to white spot syndrome virus infection. *PLoS ONE* 8(10). doi:10.1371/journal.pone.0076718.

Yang H, Wei X, Wang R, Zeng L, Yang Y, Huang G, Shafique L, Ma H, Min lv, Ruan Z, et al. (2020) Transcriptomics of *Cherax quadricarinatus* hepatopancreas during infection with Decapod iridescent virus 1 (DIV1). *Fish and Shellfish Immunology* 98: 832–842. doi:10.1016/j.fsi.2019.11.041.

Yu Y, Wei J, Zhang X, Liu J, Liu C, Li F, Xiang J (2014) SNP discovery in the transcriptome of white pacific shrimp *Litopenaeus vannamei* by next generation sequencing. *PLoS ONE* 9(1): 1–9. doi:10.1371/journal.pone.0087218.

Yu Y, Zhang X, Yuan J, Li F, Chen X, Zhao Y, Huang L, Zheng H, Xiang J (2015) Genome survey and high-density genetic map construction provide genomic and genetic resources for the Pacific white shrimp *Litopenaeus vannamei*. *Scientific Reports* 5: 1–14. doi:10.1038/srep15612.

Zeng D, Chen Xiuli, Xie D, Zhao Y, Yang C, Li Y, Ma N, Peng M, Yang Q, Liao Z, et al. (2013) Transcriptome analysis of Pacific white shrimp (*Litopenaeus vannamei*) hepatopancreas in response to Taura syndrome virus (TSV) experimental infection. *PLoS ONE* 8(2): 1–8. doi:10.1371/journal.pone.0057515.

Zhan S, Aweya JJ, Wang F, Yao D, Zhong M, Chen J, Li S, Zhang Y (2019) Litopenaeus Vannamei Attenuates White Spot Syndrome Virus Replication by Specific Antiviral Peptides Generated from Hemocyanin. *Developmental and Comparative Immunology* 91: 50–61. Elsevier Ltd.

Zhang X, Pan L, Yu J, Huang H (2019) One recombinant C-type lectin (LvLec) from white shrimp *Litopenaeus vannamei* affected the haemocyte immune response *in vitro*. *Fish and Shellfish Immunology* 89: 35–42. doi:10.1016/j.fsi.2019.03.029.

Zhong S, Mao Y, Wang J, Liu M, Zhang M, Su Y (2017) Transcriptome analysis of Kuruma shrimp (*Marsupenaeus japonicus*) hepatopancreas in response to white spot syndrome virus (WSSV) under experimental infection. *Fish and Shellfish Immunology* 70: 710–719. doi:10.1016/j.fsi.2017.09.054.

Zhu F, Wang Z, Sun BZ (2016) Differential expression of microRNAs in shrimp *Marsupenaeus japonicus* in response to *Vibrio alginolyticus* infection. *Developmental and Comparative Immunology* 55: 76–79. doi:10.1016/j.dci.2015.10.012.

Zhu G, Li S, Wu J, Li F, Zhao XM (2019) Identification of functional gene modules associated with STAT-mediated antiviral responses to white spot syndrome virus in shrimp. *Frontiers in Physiology* 10: 1–9. doi:10.3389/fphys.2019.00212.

3

Fish Transcriptomics: Applied to Our Understanding of Aquaculture

Joseph Heras

California State University, San Bernardino, California

CONTENTS

3.1 Introduction

As the demand for aquaculture products increases, the use of innovative methods to improve the health and quality of these industries are crucial. By 2050, the human population will likely reach 9 billion, which will require an increased production of food and nutrients, and will include aquaculture industries (Barange et al., 2018; World Bank, 2013). With this increasing demand in aquaculture and fisheries, we must develop efficient and productive methods to operate these industries. In addition, a vast amount of changes to environmental conditions due to climate change, along with a finite amount of resources available, poses challenges for improving aquaculture production (Barange et al., 2018). With the advancement in genetic sequencing technologies (i.e. next-generation sequencing, NGS) and computational methods, we can use genomic and transcriptomic information to better understand the biology of these aquatic sources of nutrition and improve production. Also, advancements in technology and bioinformatic tools have bolstered many "omic" fields, which include transcriptomics but also genomics, metabolomics, proteomics and metagenomics (Chandhini & Rejish Kumar, 2019). All "omic" fields including transcriptomics provide extensive amounts of data and the opportunity to understand aquaculture from a holistic approach and to address the issues associated with growth, nutrition, disease, and conservation. Aquaculture practices are global efforts, which have an economic impact and provide food resources throughout the world. Here lies many aspects of what is gained from sequencing transcriptomes from various species of fishes and how this information is applied toward the improvement of aquaculture.

With the advent of RNA-seq techniques, we can provide a better understanding of biological systems in which many studies have generated data for aquaculture purposes and have made strides and efforts on a better understanding of nutrition, growth, reproduction, immune function and stress, and adaptations for species used in aquaculture (Escalante-Rojas et al., 2018; Leduc et al., 2018; Yang et al., 2018; Table 3.1). Although many species may have similar biochemical and physiological properties, life histories, diet, and physiological adaptations, these features can be species-specific. Therefore, the use of RNA-seq techniques can be applied to better understand the physiology, behavior, evolution, endocrinology, ecology, population genetics, and conservation of the species of interests used in aquaculture. We can use these novel genetic methods to make stronger efforts in aquaculture production and conservation, in which optimal production can be available, and with a stronger understanding about the biology of these farmed species, we can strive toward stronger sustainable practices.

Transcriptomics allows us to identify transcripts (RNA molecules) from a given cell or tissue type from an organism (Wang et al., 2009). Most transcriptomic studies have focused on messenger RNA (mRNA) molecules, because these molecules demonstrate that a gene is expressed, and the mRNA molecule will be translated into a protein product which serves a function within the cell or tissue. With high-throughput sequencing technologies, such as NGS, these sequencing technologies can be used for genomic and transcriptomic applications. High-throughput transcriptomics would not be possible without the use of methods such as RNA-seq technologies (Wang et al., 2009) and the advancement in computation power and programming. With a population of RNA, whether it is total RNA or targeted with poly(A)+, these molecules are transformed into a library of cDNA fragments (Wang et al., 2009). With these libraries, next-generation sequencing (i.e. Illumina, Oxford Nanopore, and Pacific Biosciences) can be implemented to generate a vast amount of sequence reads, which yields over a Gigabase (GB) (10^9 nucleotides). In the past, transcriptomes were quantified with either sequence or

TABLE 3.1

Contemporary Papers on Aquaculture Transcriptomics (2017–2020)

Study Focus	Species	Common Name	Tissue Type(s)	References
Adaptive evolution	*Sebastes carnatus, S. nebulosus, S. maliger, S. mystinus,* and *S. serranoides*	Gopher, China, quillback, blue, and olive rockfishes	Brain	Heras and Aguilar (2019)
Climate change-elevated pCO_2	*Sebastes carnatus* and *S. mystinus*	Gopher and blue rockfishes	White muscle	Hamilton et al. (2017)
Climate change stress	*Sebastes mystinus*	Blue rockfish	White muscle, gill, and liver	Cline et al. (2020)
Diet	*Dicentrarchus labrax*	European sea bass	Proximal intestines	Leduc et al. (2018)
Diet and immune response	*Epinephelus akaara*	Hong Kong grouper	Liver	Yang et al. (2018)
Diet replacement	*Salmo salar* L.	Atlantic salmon	Liver, midgut, and hindgut	Betancor et al. (2017)
Egg quality	*Dicentrarchus labrax*	Sea bass	Fertilized eggs	Żarski et al. (2017)
Fishmeal based diet	*Carassius gibelio*	Gibel carp	Liver	Xu et al. (2019)
Genome and transcriptome characterization of the digestive tract	*Seriola quinqueradiata*	Japanese amberjack (yellowtail)	Gills, skin, fins, red muscle, white muscle, heart, kidney, spleen, stomach, intestine, pyloric caeca, liver, gallbladder, retina, cerebellum, optic lobe, olfactory lobe, ovary	Yasuike et al. (2018)
Growth/atrophy	*Lutjanus guttatus*	Rose spotted snapper	Spleen, gills, brain, heart, testis, liver, gut, muscle, dark skin, white skin and visceral fat	Escalante-Rojas et al. (2018)
Head shape dimorphism/ growth rate and chemotaxis	*Anguilla anguilla*	European glass eel	Head	Meyer et al. (2017)

(*Continued*)

TABLE 3.1 *(Continued)*

Contemporary Papers on Aquaculture Transcriptomics (2017–2020)

Study Focus	Species	Common Name	Tissue Type(s)	References
Nutrition (plant-based feed)	*Salmo salar*	Atlantic salmon	Liver	Vera et al. (2017)
Sex-related differences	*Acipenser schrenckii*	Amur sturgeon	Testes, ovaries, and livers	Zhang et al. (2019)
Sex-related/sex determination	*Oreochromis niloticus*	Nile tilapia	Gonads	Tao et al. (2018)
Sex-related/sex determination	*Symphysodon haraldi*	Discus fish	Gonads	Lin et al. (2017)
Sex reversal/ temperature	*Oreochromis niloticus*	Nile tilapia	Gonads	Wang et al. (2019)
Stress response and aggression	*Oreochromis niloticus*	Nile tilapia	Hypothalamus	Rodriguez-Barreto et al. (2019)

hybridization on custom-made microarrays (Clark et al., 2002; Reinartz et al., 2002), but this required prior knowledge of genomic sequences and were time-consuming and highly laborious (Chandhini & Rejish Kumar, 2019). With current methods, prior genomic information is not required to sequence a transcriptome of any organism. Once the transcriptome has been produced from one of the various sequencing platforms, multiple analyses can be conducted to give insight on the cellular processes for any species of interest and investigate questions associated with changes in diet, growth and development, exposure to different environmental conditions, or exposure to a pathogen. Depending on the objectives, transcriptomics can target specific cells or tissues.

Analyses can include transcriptome annotation, SNP detection, differential gene expression, gene enrichment, detection of alternative splice variants, identification of orthologous sequences, and estimation of positive Darwinian selection. It is evident the abundant amount of studies applying RNA-seq techniques to aquaculture aimed at improving the health and quality of the desired species of interest (Chandhini & Rejish Kumar, 2019).

Also, the transcriptomic information has been a cost-efficient tool to utilize, especially in the absence of a genome for the species of interests. However, with the constant reduction in sequencing costs, genomes can be readily available and used as a complement to transcriptomic analyses. In the past couple of years, an extensive amount of research was done to develop fast and reliable software needed for genomic sequencing on "single node" computers which contain multi-threaded processors (Martinez et al., 2018). The transcriptome can be used in a wide range of applications such as identifying a collection of protein-encoding genes, the composition of a gene with alternative splicing and post-translation modifications, and lastly, measure of differential expression of hundreds to thousands of genes within a given individual (Chandhini & Rejish Kumar, 2019). Transcriptomic studies allow us to better understand gene expression within a temporal framework, where we can ask questions about how gene expression changes during different time periods. This insight on any changes in gene expression profiles can better explain which genes are involved in growth, diet shifts, metabolism, and development. This temporal aspect used in transcriptomic analyses cannot be identified solely with the use of genomic sequencing. Genomic sequencing provides only the location of genes within chromosomal arrangements; it does not provide insight on how these genes are expressed within a given tissue or time point. Many studies investigate which genes are up-regulated or down-regulated during developmental stages, which is crucial in understanding growth. With improved insight on growth and reproduction, many industries seek to produce quick growth rates and simultaneously reduce the amount of stress and diseases present during production. Gene expression patterns provide a unique opportunity to better understand how organisms respond to altered environmental conditions such as co-occurrence of hypoxia and high pCO_2 (Cline et al., 2020) or exposure to pollutants (Pojolar et al., 2012). Also, transcriptomic analyses can have a spatial context (e.g. targeting a specific tissue type or whole organism). These questions can provide answers about genes which encode for proteins within a given tissue that are essential for a physiological function such as digestion, reproduction, and/or adaptations.

As more genomic resources become available for more species with importance for aquaculture, many studies have focused on the use of model species such as zebrafish (*Danio rerio*), Japanese rice fish (*Oryzias latipes*) and Japanese pufferfish (*Fugu rubripes*), which have sequenced genomes and are used to make inferences about growth, development, fish physiology, diet, metabolism, immune function and response to toxins (Mazurais et al., 2011; Ulloa et al., 2014). Many bioinformatic pipelines have been developed with the genome and transcriptomes from these model species, and they have laid the methodology for studies on non-model organisms. Many studies have shown multiple genes that are up-regulated and down-regulated in which multiple experimental designs can be developed to understand a variety of aspects of an organism such as metabolism of nutrients, disease and stress pathways, and development of specific organs, and the results of these studies can be applied to aquaculture (Dahm & Geisler 2006). As more transcriptomic and genomic studies are generated, we are able to delineate patterns that transcend multiple species or can be restricted to a few species because of their unique evolutionary adaptations.

Currently, a vast amount of transcriptomic and genomic sequence data is being generated for multiple organisms, where the data extensively exceed the amount of analyses completed (Schadt et al., 2010). This immense amount of sequence data available requires us to adopt the use of advanced computational resources (Schadt et al., 2010). The data are deposited on major sequence databases (DNA Data Bank of Japan—DDBJ [Japan], GenBank [USA], and European Nucleotide Archive [Europe]). This information can be used for many comparative genomic and transcriptomic analyses. In addition, within the past 20 years sequencing data costs have decreased dramatically, where the cost per megabase (Mb or 1,000,000 base pairs) of sequence data is a few U.S. cents (Wadapurkar & Vyas 2018). The cheap costs and availability of genetic sequencing gives rise for efforts to obtain sequencing information for the species of interests. It also allows analysis of closely related species which can give a broader understanding about the ecological and evolutionary relationships which can be crucial for conservation of species and aquaculture management. Aquaculture genetics had begun around 80 years ago and has steadily grown in the past 30 years (Dunham & Lucas 2019). Advancements in genetics and breeding research has been established for commercial species such as carps, catfish, salmonids, tilapias, and oysters. All these species have been genetically improved (Dunham & Lucas 2019). However, genetic studies on aquaculture species are still scarce, as some reports show that only 10% of commercial aquaculture species are genetically improved (Gjedrem & Robinson 2014; Gjedrem et al., 2012).

The vast amount of genetic information generated prompts to ask the following questions: What type of immune response occurs in the presence of a pathogen? Which genes are expressed when the composition of a diet has been changed (i.e. less fish meal)? How does this novel diet or lack of nutrients impact immune function, growth and reproduction? How has this bulk of information improved our management for conservation and fisheries? New methods will be crucial for developing demand of the aquaculture in the upcoming years. The present chapter deals with the application of transcriptomics and its vital importance in the understanding of fisheries and aquaculture.

3.2 Nutrition, Growth, and Development

3.2.1 Nutrition and Growth

One of the most crucial objectives in aquaculture is selecting nutrient resources optimal for aquaculture. However, what impacts do these diets have on the species in aspects such as growth, digestion, metabolism, immune function, and reproduction? Certainly, many factors play into the health and quality of the species of interest, but most questions revolve around the type of nutrients that can be given with the lowest cost but provide the highest yield in growth of fishes or invertebrates. This goal is ubiquitous for any farming industry to create the fastest yield; however, compromises come in the form of changing the diet can shift the quality and/or taste of the product (Calanche et al., 2020). In the past, fish meals (FM), mostly made from fish and not intended for direct human consumption, have been generated for carnivorous fishes, which contains a high amount of protein, with essential amino acids and fats (Hardy, 2010). Also, this provides a dilemma, in which the availability of wild fish for the production of fish meal for any aquaculture species is limited, and the aquaculture production becomes a net consumer as opposed to a net producer (Hardy, 2010). It has been a focus on fish feed and fish oil (FO) as a feed source for aquaculture. However, a shift has occurred giving the basic nutrients for aquaculture, in which "functional feeds" have been designed to improve the health and growth of a fish (Tacchi et al., 2011). At present, additives are included in feeds and they comprise prebiotics, probiotics, immunostimulants, vitamins, nucleotides, minerals, and plant and algal extracts. The efficacy of these new strategies, shifting the diet of fishes, can be evaluated through transcriptomic profiles in order to understand how this may alter the physiological conditions of tissues within aquaculture species, and how this may benefit or be a detriment to the production of the species of interests, particularly in the digestive tract.

The gastrointestinal tract offers a great insight on many functions such as digestion, osmoregulation, detoxification, or immune function (Karasov & Douglas, 2013). Many studies have monitored the gut of fishes when presenting some type of modification within their diet (Calduch-Giner et al., 2016; Kaushik, 2002; Lu et al., 2019; Morais et al., 2012). Several aquaculture species are carnivorous, which require a large protein intake. For instance, the European seabass (*Dicentrarchus labrax*) requires a vast amount of animal protein (43%–50% of its daily food intake) (Kaushik, 2002). In addition to proteins, carbohydrates, and lipid content within fish feed, the function feed is crucial in which it serves health benefits for the fish (Martin & Król, 2017). The function fish feed includes selenium, zinc, and vitamins and may contain other ingredients such as immunostimulants in which algal and plant extracts are integrated into these feeds. Also, prebiotics, which usually stem from yeast extracts in order to stimulate the microbiome operating within the fish intestine (Hoseinifar et al., 2016). In all the modifications of diet within aquaculture, so much insight is to be gained and can improve the quality of the aquaculture species of interest.

The ingredients FM and FO have been traditional marine resources as feed and a major source of dietary protein and lipids for aquaculture. Recently, it has been a transition to include terrestrial plant products such as plant meals and vegetable oils (VO) into fish feed (Hardy, 2010; Naylor et al., 2009). This shift in diet for aquaculture purposes is critical to reduce the reliability of fish meal or wild-caught fish used for fish feed which is a finite and ever more expensive resource (Naylor et al., 2009). Many farmed fish species require a carnivorous diet, and now with the increasing use of plant meal and VO in feed, questions arise on whether this shift to a plant protein-based diet has impacts on growth, immunity, and digestive physiology. Particularly for questions regarding absorption and digestion, the use of gut transcriptomics can elucidate a better understanding of metabolism and digestion. It is evident that the type of fish feed used in aquaculture impacts multiple physiological aspects such as nutrition, development and health condition of the fish (Tacchi et al., 2011). Studies have shown that alternative plant-based ingredients, as opposed to marine derived products, can have negative impacts on fish physiology such as growth, health, and disease resistance (Hardy, 2010). Nutrigenomics is a recent discipline that serves to improve many industries including aquaculture. It has focused on the molecular movement and other dietary components required for aquaculture species (Hakim et al., 2018; Vera et al., 2017). This field focuses on developing feed, which has positive effects of economic value, the welfare of the fish, and supplying market demands (Hakim et al., 2018). These studies involve altered diet conditions, and their effects are evaluated at the transcriptomic, proteomic, metabolomic, and epigenomic levels (Hakim et al., 2018; Vera et al., 2017). It is clear that the use of transcriptomics along with other fields of biology can provide profound answers to many of the questions posed concerning fish aquaculture. Studies done on the rose spotted snapper (*Lutjanus guttatus*) when given a vegetable-based diet recorded negative impacts where a vegetable diet did not fulfill all the amino acid requirements for a developing fish (juvenile). Also, a decrease in gene expression of myogenic and other growth-related genes was recorded (Escalante-Rojas et al., 2018). Transcriptomic studies can be used to show the effects of diet modifications, especially in supplementing a plant-based diet to carnivorous or omnivorous species. These type of studies are necessary to understand how the species of interest responds to this altered diet at the molecular level. Ingredients such as soybean meal (SBM), soybean protein concentrate (SPC), corn gluten, sunflower meal, and pea protein concentrate (PPC) are commonly used in fish feed (Vera et al., 2017). It has been shown that these plant ingredients contain secondary metabolites, which can be anti-nutritional factors (ANFs). These metabolites can reduce feed intake and impair the nutrients that are within vegetable-based feeds (Vera et al., 2017). When a plant-based diet is introduced, a shift from a marine diet containing fish meal (80%) to a vegetable diet (10% FM), has had an impact on the Atlantic salmon (*Salmo salar*), showing up-regulated genes involved in oxidative phosphorylation, pyruvate metabolism, tricarboxylic acid (TCA) cycle, glycolysis, and fatty acid metabolism (Vera et al., 2017), as shown in the liver transcriptome. An additional factor that was considered in such a study was ploidy (i.e. 2n and 3n) which could be a potential factor influencing how a diet shift may impact the development of these fishes. Therefore, different nutritional requirements have been shown for triploid salmon as compared to diploid salmon (Vera et al., 2017).

Extensive studies have been done on the digestive physiology of fishes which were applied to many facets of biology, including applications to aquaculture (De Santis et al., 2015; German et al., 2015; German et al., 2016; Wang et al., 2010). A better understanding of the digestive enzymatic activity during a shift in diet can provide a better understanding of what type of activity is operating within the fishes of interest. Many digestive enzymes play a crucial role in digestion and absorption, but the gene activation of such enzymes are dependent on the diet of the species of interest. In addition, many aquaculture species may have a broad range of diets (i.e. carnivorous, omnivorous, or herbivorous) or even contain an ontogenetic diet shift through their life history (Chinook salmon; Duffy et al., 2010). Transcriptomics may help to get a better understanding of the optimal enzymatic activity depending on the species and its feeding habits. There are two compelling hypotheses. First is the Adaptive Modulation Hypothesis (AMH; Karasov, 1992), which states a positive correlation between the digestive enzyme activities and the substrate that has been ingested that corresponds to those enzymes. This has been supported in many studies focused on carbohydrases from both herbivorous and omnivorous animals. In contrast, the Nutrient Balancing (NB) Hypothesis (Clissold et al., 2010) supports the elevated expression levels of enzymes for a limiting dietary resource to ensure acquisition of these nutrients such as essential fatty acids. The application of gut transcriptomics in these issues may contribute to a better understanding of which genes are expressed associated with digestive enzymes, and may help to prove which of these two hypotheses are likely to occur in some species.

The efficient formulation of feed can be developed for a specific fish species by understanding which proteolytic enzymes are necessary for its digestion and absorption (Yasuike et al., 2018). A study done on the Japanese amberjack (*Seriola quinqueradiata*) highlighted the expression of genes coding for proteolytic digestive enzymes such as trypsinogen, two chymotrypsinogen genes, and a carboxypeptidase B gene within the intestine and rectum. Other digestive genes found were three apolipoproteins: apolipo Eb, apolipo B-100, and apolipo A-1. These apolipoproteins are vital in lipid transport and uptake in vertebrates and have been shown to be synthesized primarily in the intestine and liver of most teleost. As a caveat, it has been suggested that the fish intestine transcriptomes are plastic, in which they can change spatially, seasonally, and with diet (Calduch-Giner et al., 2016). This provides a key detail of matching the available digestive enzymes present in the fish species with the corresponding diet presented. Giving a general diet to fish with different feeding preferences lacking the required digestive enzymes to digest and absorb the feed can be counterproductive and detrimental to their health. In order to ensure the optimal digestion/absorption of nutrients, the formulation of specialized feed and/or the use of gene analysis methods such as CRISPR/Cas9 (gene editing) can be used. In this way, it can be ensured that resources are not wasted and the health of the fish species is not compromised.

Further, given that many fish species have external fertilization and high fecundity rates, they would be good candidates for gene editing as compared to terrestrial animals (Gratacap et al., 2019). More genomic and transcriptomic studies are being produced with species that naturally have an herbivorous diet such as the monkeyface prickleback fish *Cebidichthys violaceus* (Heras et al., 2020). Although this is not a widely cultured fish, a better understanding of organisms that can digest and acquire nutrients from an herbivorous diet serves as a template for which genes are expressed and encode for proteins that aide in the digestion of plants. This is vital information that can be used as a guide to what the genetic requirements are on a plant-based diet and reduce the risk of an ANF which can lead to negative effects on the health of fishes.

3.3 Reproduction and Development

In any form of culture or breeding program, reproduction plays a crucial role in large-scale commercial industries, in which some form of artificial reproduction is taking form. When dealing with species in the wild, species identification is crucial because the possibility of hybridization can occur. In addition, many marine species have wide distributions in which a putative population structure can occur along their distribution. Depending on the species in aquaculture facilities, various forms of artificial reproduction can be done, but to what extent does this impact the progeny and subsequent generations with this form of artificial selection? Transcriptomic analyses can provide some insight on fitness of individuals and also an understanding of how reproduction at the molecular level operates within the species of interest.

An understanding of the genetic loci responsible in determination of sex in fishes can provide insight into reproduction, which is helpful for aquaculture and fisheries. Within fishes, sex determination is a dynamic process and has been observed in families or genera which can be modulated by external factors (Devlin & Nagahama, 2002). Sex determination can occur through monogenic or polygenic systems, which can be located on autosomes or sex chromosomes. With the use of transcriptomics, we can compare sexes to determine which loci contribute to sex determination (Chen et al., 2015; Lin et al., 2017; Sun et al., 2013). The transcriptome from reproductive tissue can also provide biological insight on understanding sex-related differences or sex differentiation in fishes (Tao et al., 2018; Zhang et al., 2019). Multiple candidate genes associated with sex determination, gametogenesis, and gonadal differentiation and maturation have been identified within fishes, such as Dmy/dmrt1Yb, DM-W, DMRT1, Sox9, SDy, and Sox3 (Matsuda et al., 2002; Takehana et al., 2014; Yano et al., 2012; Yokoi et al., 2002). The Amur sturgeon

(*Acipenser schrenckii*), a species used for generating caviar, was used in a transcriptomic study that showed how long non-coding RNAs (lncRNAs) might be one factor in regulating differential gene expression associated with sex-related differences (Zhang et al., 2019). More transcriptomic studies aimed at sex determination in fishes are still necessary in order to make broader inferences about reproduction, which can be helpful for aquaculture management.

For both wild and domestic animals, egg production can be impacted by many factors, some of which are highly variable, where the production of inviable eggs may be common for a variety of species (Chapman et al., 2014). At early stages of vertebrate embryo development, rapid cell division processes occur in a synchronous fashion where cells are dividing in the zygote to form a blastula. In this stage, maternal RNAs are essential, in which these gene transcripts were inherited in growing oocytes and direct embryogenesis (Chapman et al., 2014). In such studies, these transcripts encode regulators or participants in cell cycling (cyclins, nucleoplasmin), proliferation, growth and apoptosis (insulin-like growth factors and their receptor, prohibitin) and cytoskeleton (tubulin beta, keratins 8 and 18; Aegerter et al., 2005; Bonnet et al., 2007a; Bonnet et al., 2007b). Farmed striped bass has had issues with egg quality (Chapman et al., 2014). Studies done on zebrafish and Japanese rice fish (medaka) have provided insights on reproduction and development (Qiao et al., 2016; Vesterlund et al., 2011), that can offer a better understanding of these issues. For instance, the use of gene ontology (GO) annotations from *D. rerio* were used in a gene set enrichment analysis (GSEA) with head tissue from European glass eels to better understand the development of head shape dimorphisms (De Meyer et al., 2017). As more genomic and transcriptomic data becomes more readily available, we can make stronger inferences about fertilization, sex determination, and reproductive physiology in fishes. This insight can be applied to making better decisions in spawning management.

3.4 Immune Function, Stress, and Toxicology

Understanding which genes are expressed in the presence of a pathogen and the diversity of genes related to immune function are important aspects for aquaculture and fisheries. Multiple immune function genes are known to be under positive Darwinian selection, which is key for maintaining a healthy population within the confinement of a facility (Xiao et al., 2015). Multiple studies in vertebrates show that Major Histocompatibility genes are under positive Darwinian selection, in which they are part of the adaptive immune system (Aguilar & Garza, 2005; Schad et al., 2012; Xiao et al., 2015). Candidate tissues for transcriptomic studies would be head kidney, kidney, and spleen for understanding immune function (Chandhini & Rejish Kumar, 2019). Immune function does not operate in isolation, in which nutrition also plays a role in sustaining health of aquaculture species (Martin et al., 2010). It was shown that fasting Atlantic salmon (*Salmo salar*) decreased expression of immune function within the liver in the presence of a bacterial pathogen as compared to fishes that were fed and exposed to the pathogen (Martin et al., 2010). This suggests a mechanism for conservation of energy during starvation. Therefore, any alterations in feed can also have positive or negative impacts on immune function. If negative impacts exist on immune function, this provides stress on the organism which can result in a reduction of longevity.

Environmental stress can occur in multiple forms, such as differences in temperature, salinity, pH, spatial constraints, and pollutants. Exposure to different environmental conditions that provides some form of stress on the aquaculture species could hinder growth, development, and overall health of the fish. Several studies showed that stress exposure elicits a physiological and biochemical response within teleost fishes (Aluru & Vijayan, 2009). Zebrafishes and medaka provide great models for understanding how pollutants can impact certain biological aspects. Harmful consequences have been reported in development, endocrine, and reproductive aspects (Weber et al., 2013). A better understanding of the immune function and exposure to specific pollutants or stressors can have an extensive impact of growth, reproduction, and longevity. The use of transcriptomics holds strong promise in extending the lives of individuals within aquaculture and also can be a litmus test for ensuring the health of these species, which is important for sustaining aquaculture optimal for human consumption as well.

3.5 Adaptations, Molecular Evolution, and Population Genetics

It is clear that physiological adaptations are crucial for organisms to survive within their habitats, especially if human activities have altered these conditions, due to aquapens for mass production or as a consequence of climate change impacting the temperature of the oceans. Many species of economic value can also serve with great interests to applications in evolution. Multiple studies have used comparative transcriptomics to identify orthologous gene pairs/clusters and then estimate natural selection by using a form of estimating non-synonymous and synonymous rates (omega), where an omega value > 1 indicates a positive selection, a value = 1 means a neutral selection, and a value < 1 is purifying selection (Heras & Aguilar 2019; Heras et al., 2011; Heras et al., 2015; Tong et al., 2015; Xiao et al., 2015). Many of these genes can provide us with a better understanding if general patterns of selection are operating on certain genes. This is especially true if multiple species are exposed to similar habitats (e.g. temperature, pH, oceanic depth, differences in salinity). It is clear that some genes under positive Darwinian selection contribute to a physiological adaptation in a given environment. Marine rockfishes (*Sebastes*) serve this unique opportunity to understand evolution of a genus that is composed of 105 species (Love et al., 2002) but is also important when applied to our understanding to aquaculture purposes. Rockfishes are clear examples of adaptive radiations within marine habitats, and they are a cluster of multiple species which can live more than 100 years. Multiple candidate genes under positive Darwinian selection have been proposed (Heras & Aguilar, 2019; Heras et al., 2011; Heras et al., 2015). These estimates of positive Darwinian selection provide some insight on adaptation at the molecular level and elucidate patterns within this speciose group. Several candidate genes of interests involving genes associated with metabolism, longevity, reproduction, and oxygen consumption have been described. It was shown in Heras and Aguilar (2019) that hemoglobin subunit alpha is under strong positive Darwinian selection within rockfishes through a comparative transcriptomic analysis. In addition, it enhances our understanding of how well a species responds to novel environmental conditions such as changes in pCO_2, which is likely in the occurrence of climate change within the oceans. A study done by Hamilton et al. (2017) with rockfishes has shown their transcriptomic profile in the presence of elevated pCO_2, a simulated environment due to anticipated climate change. Studies like these are of interest because not only do we understand the evolutionary adaptation of the species of interest, but also its meaning for efforts in conservation and fisheries.

The fish-specific genome duplication (FSGD) has been estimated to have occurred around ~350 million years ago (Meyer & Van de Peer, 2005). The presence of a duplicated genome provides the opportunity for novel gene functions to occur via neofunctionalization or subfunctionalization (Wolfe, 2001). With novel functions, these duplicated genes can also be favorable and can become fixed within a population as a result of positive selection. About ~25,000 different species of teleost fishes are presently known, more than any other vertebrate group (Meyer & Van de Peer 2005). Therefore, the presence of gene duplications gives the opportunity for expression profiles to be different from what is seen in other vertebrates. Although many facets exist that contribute to adaptation and speciation, some of the basic understanding of the evolutionary patterns within organisms can be assessed with access to genomic information such as the identification of positive selection within the transcriptome and the identification of orthologous and paralogous genes. In addition, understanding which loci are expressed, or whether multiple paralogous genes are expressed, can provide insight on using gene-editing methods to make sure that similar copies are used for a desired phenotype in the species of interest.

Population genetics or genomic studies have always played a strong role in understanding population structure of a variety of aquaculture species such as salmon, trout, and rockfishes. Although most population genetic studies focus on genomic sequencing information to detect SNPs, this information gives a better understanding about population structure or identifying candidate loci responsible for local adaptive traits (Nielsen et al., 2009). Transcriptomic information can be used to make these inferences about divergence in populations, which has been conducted between dwarf and normal whitefish to identify SNP markers within coding regions (Renaut & Bernatchez, 2011). With more loci to analyze, population genomic analyses can give a better understanding about the evolutionary processes within or between fishes that are reared in captivity or in the wild.

3.6 Microbiome and Applications for Conservation and Aquaculture Management

When thinking about the biology of a given aquaculture species and if an immense amount of transcriptomic data was available for such species, there is still more biological information that can be insightful since an organism does not operate in isolation. Many bacteria species are commonly viewed as pathogens, but many others have mutualistic relationships with fish that translate as health of a fish. A systems biology approach already exists in the form of a Holo-Omic effort to represent a broad understanding of the biology of a species of interest, which includes the microbiome (Limborg et al., 2018). With the current technological advances and knowledge advancement of microbial genomes, pairing the transcriptome of bacteria residing within fish of interest, and aids in digestion or other physiological functions, will provide greater insight on genomic and transcriptomic studies. This approach offers many opportunities to better understand co-evolutionary processes that operate between the gut microbiome and the host (Mekuchi et al., 2019). In addition, metatranscriptomics (transcriptomes of microbiota) serves as a useful tool for understanding how microbial genes aid in digestion (Wu et al., 2015).

The use of transcriptomics, in a subset of genes expressed within the genome, is a powerful resource for rapidly identifying coding genes and noncoding transcripts from the genome as opposed to sequencing the genome with no prior knowledge. Further, many studies contributed to advance in knowledge of the constituents of a genome which can provide understanding on loci that are associated with risk factors for genetic disorders, variation in immune function, specialization (i.e. dietary tolerances), and reproduction (Ryder, 2005). These factors may provide insightful information not only for evolutionary biology but for aquaculture, fisheries, and conservation management as well. We can use transcriptomics to advance our understanding of commercially important fishes and reveal how biological processes are operating at the transcriptomic level. Many species are overfished, and some have been listed either as threatened or as endangered species (Magnuson-Ford et al., 2009). The advancement in transcriptomic studies can elucidate our understanding of risk factors for genetic disorders, variation in immune function, specialization (i.e. dietary tolerances), and reproduction. These can enhance our knowledge about physiology and to propose an effective guide for conservation management by means of improving wildlife health conditions and plans for intervention of population viability (Ryder, 2005). With recent advancements in genetic sequencing technology and computational capabilities, aquaculture can further advance in various fields including health of the farmed species, which in turn will benefit the human.

References

Aegerter S, Jalabert B, Bobe J (2005). Large scale real-time PCR analysis of mRNA abundance in rainbow trout eggs in relationship with egg quality and post-ovulatory ageing. *Molecular Reproduction and Development: Incorporating Gamete Research.* 72(3): 377–385.

Aguilar A, Garza JC (2005). Analysis of major histocompatibility complex class II Beta genes from rockfishes (genus *Sebastes*). *Journal of Fish Biology.* 67(4): 1021–1028.

Aluru N, Vijayan MM (2009). Stress transcriptomics in fish: A role for genomic cortisol signaling. *General and Comparative Endocrinology.* 164(2–3): 142–150.

Barange M, Bahri T, Beveridge MC, Cochrane KL, Funge-Smith S, Poulain F (2018). *Impacts of Climate Change on Fisheries and Aquaculture. Synthesis of Current Knowledge, Adaptation and Mitigation Options.* Rome: Food and Agriculture Organization of the United Nations.

Betancor MB, Li K, Sprague M, Bardal T, Sayanova O, Usher S, … Tocher DR (2017). An oil containing EPA and DHA from transgenic *Camelina sativa* to replace marine fish oil in feeds for Atlantic salmon (*Salmo salar* L.): Effects on intestinal transcriptome, histology, tissue fatty acid profiles and plasma biochemistry. *PloS ONE.* 12(4): e0175415.

Bonnet E, Fostier A, Bobe J (2007a). Microarray-based analysis of fish egg quality after natural or controlled ovulation. *BMC Genomics.* 8(1): 55.

Bonnet E, Montfort J, Esquerre D, Hugot K, Fostier A, Bobe J (2007b). Effect of photoperiod manipulation on rainbow trout (*Oncorhynchus mykiss*) egg quality: A genomic study. *Aquaculture.* 268(1–4): 13–22.

Calanche JB, Beltrán JA, Hernández Arias AJ (2020). Aquaculture and sensometrics: The need to evaluate sensory attributes and the consumers' preferences. *Reviews in Aquaculture.* 12(2): 805–821.

Calduch-Giner JA, Sitjà-Bobadilla A, Pérez-Sánchez J (2016). Gene expression profiling reveals functional specialization along the intestinal tract of a carnivorous teleostean fish (*Dicentrarchus labrax*). *Frontiers in Physiology*. 7: 359.

Chandhini S, Rejish Kumar VJ (2019). Transcriptomics in aquaculture: Current status and applications. *Reviews in Aquaculture*. 11(4): 1379–1397.

Chapman RW, Reading BJ, Sullivan CV (2014). Ovary transcriptome profiling via artificial intelligence reveals a transcriptomic fingerprint predicting egg quality in striped bass, *Morone saxatilis*. *PLoS ONE*. 9(5): e96818.

Chen X, Mei J, Wu J, Jing J, Ma W, Zhang J, … Gui JF (2015). A comprehensive transcriptome provides candidate genes for sex determination/differentiation and SSR/SNP markers in yellow catfish. *Marine Biotechnology*. 17(2): 190–198.

Clark TA, Sugnet CW, Ares M (2002). Genomewide analysis of mRNA processing in yeast using splicing-specific microarrays. *Science*. 296(5569): 907–910.

Cline AJ, Hamilton SL, Logan CA (2020). Effects of multiple climate change stressors on gene expression in blue rockfish (*Sebastes mystinus*). *Comparative Biochemistry and Physiology Part A: Molecular & Integrative Physiology*. 239: 110580.

Clissold FJ, Tedder BJ, Conigrave AD, Simpson SJ (2010). The gastrointestinal tract as a nutrient-balancing organ. *Proceedings of the Royal Society B: Biological Sciences*. 277(1688): 1751–1759.

Dahm R, Geisler R (2006). Learning from small fry: The zebrafish as a genetic model organism for aquaculture fish species. *Marine Biotechnology*. 8(4): 329–345.

De Meyer J, Maes GE, Dirks RP, Adriaens D (2017). Differential gene expression in narrow-and broad-headed European glass eels (*Anguilla anguilla*) points to a transcriptomic link of head shape dimorphism with growth rate and chemotaxis. *Molecular Ecology*. 26(15): 3943–3953.

De Santis C, Bartie KL, Olsen RE, Taggart JB, Tocher DR (2015). Nutrigenomic profiling of transcriptional processes affected in liver and distal intestine in response to a soybean meal-induced nutritional stress in Atlantic salmon (*Salmo salar*). *Comparative Biochemistry and Physiology Part D: Genomics and Proteomics*. 15: 1–11.

Devlin RH, Nagahama Y (2002). Sex determination and sex differentiation in fish: An overview of genetic, physiological, and environmental influences. *Aquaculture*. 208(3–4): 191–364.

Duffy EJ, Beauchamp DA, Sweeting RM, Beamish RJ, Brennan JS (2010). Ontogenetic diet shifts of juvenile Chinook salmon in nearshore and offshore habitats of Puget Sound. *Transactions of the American Fisheries Society*. 139(3): 803–823.

Dunham R (2019). Genetics. In: Lucas JS, Southgate PC, Tucker CS (eds), Aquaculture: Farming aquatic animals and plants. West Sussex, UK: Wiley-Blackwell.

Escalante-Rojas M, Peña E, Hernández C, Llera-Herrera R, Garcia-Gasca A (2018). *De novo* transcriptome assembly for the rose spotted snapper *Lutjanus guttatus* and expression analysis of growth/atrophy-related genes. *Aquaculture Research*. 49(4): 1709–1722.

German DP, Foti DM, Heras J, Amerkhanian H, Lockwood BL (2016). Elevated gene copy number does not always explain elevated amylase activities in fishes. *Physiological and Biochemical Zoology*. 89(4): 277–293.

German DP, Sung A, Jhaveri P, Agnihotri R (2015). More than one way to be an herbivore: Convergent evolution of herbivory using different digestive strategies in pricklebacks fishes (Stichaeidae). *Zoology*. 118(3): 161–170.

Gjedrem T, Robinson N (2014). Advances by selective breeding for aquatic species: A review. *Agricultural Sciences.* 5(12): 1152.

Gjedrem T, Robinson N, Rye M (2012). The importance of selective breeding in aquaculture to meet future demands for animal protein: A review. *Aquaculture.* 350: 117–129.

Gratacap RL, Wargelius A, Edvardsen RB, Houston RD (2019). Potential of genome editing to improve aquaculture breeding and production. *Trends in Genetics.* 35(9): 672–684.

Hakim MM, Ganai NA, Ahmad SM, Asmi O, Akram T, Hussain S, ... Gora AH (2018). Nutrigenomics: Omics approach in aquaculture research to mitigate the deficits in conventional nutritional practices. *Journal of Entomology and Zoology Studies.* 6(4): 582–587.

Hamilton SL, Logan CA, Fennie HW, Sogard SM, Barry JP, Makukhov AD, Bernardi G (2017). Species-specific responses of juvenile rockfish to elevated pCO_2: From behavior to genomics. *PloS ONE.* 12(1): e0169670.

Hardy RW (2010). Utilization of plant proteins in fish diets: Effects of global demand and supplies of fishmeal. *Aquaculture Research.* 41(5): 770–776.

Heras J, Aguilar A (2019). Comparative transcriptomics reveals patterns of adaptive evolution associated with depth and age within marine rockfishes (*Sebastes*). *Journal of Heredity.* 110(3): 340–350.

Heras J, Chakraborty M, Emerson JJ, German DP (2020). Genomic and biochemical evidence of dietary adaptation in a marine herbivorous fish. *Proceedings of the Royal Society B.* 287(1921): 20192327.

Heras J, Koop BF, Aguilar A (2011). A transcriptomic scan for positively selected genes in two closely related marine fishes: *Sebastes caurinus* and *S. rastrelliger*. *Marine Genomics.* 4(2): 93–98.

Heras J, McClintock K, Sunagawa S, Aguilar A (2015). Gonadal transcriptomics elucidate patterns of adaptive evolution within marine rockfishes (*Sebastes*). *BMC Genomics.* 16: 656.

Hoseinifar SH, Ringø E, Shenavar Masouleh A, Esteban MÁ (2016). Probiotic, prebiotic and synbiotic supplements in sturgeon aquaculture: A review. *Reviews in Aquaculture.* 8(1): 89–102.

Karasov WH (1992). Tests of the adaptive modulation hypothesis for dietary control of intestinal nutrient transport. *American Journal of Physiology-Regulatory, Integrative and Comparative Physiology.* 263(3): R496–R502.

Karasov WH, Douglas AE (2013). Comparative digestive physiology. *Comprehensive Physiology.* 3(2): 741–783.

Kaushik SJ (2002). European sea bass, *Dicentrarchus labrax*. In: Webster CD, Lim C (eds), *Nutrient Requirements and Feeding of Finfish for Aquaculture*. Wallingford, Oxon, UK: CABI Publishing, 28–39.

Leduc A, Zatylny-Gaudin C, Robert M, Corre E, Le Corguille G, Castel H, ... Henry J (2018). Dietary aquaculture by-product hydrolysates: Impact on the transcriptomic response of the intestinal mucosa of European seabass (*Dicentrarchus labrax*) fed low fish meal diets. *BMC Genomics.* 19(1): 396.

Limborg MT, Alberdi A, Kodama M, Roggenbuck M, Kristiansen K, Gilbert MTP (2018). Applied hologenomics: Feasibility and potential in aquaculture. *Trends in Biotechnology.* 36(3): 252–264.

Lin R, Wang L, Zhao Y, Gao J, Chen Z (2017). Gonad transcriptome of discus fish (*Symphysodon haraldi*) and discovery of sex-related genes. *Aquaculture Research.* 48(12): 5993–6000.

Love MS, Yoklavich M, Thorsteinson LK (2002). *The rockfishes of the northeast Pacific.* Univ of California Press.

Lu Xing, Wen Hua, Li Qing, Wang G, Li P, Chen J, Sun Y, Yang CG, Wu F (2019). Comparative analysis of growth performance and liver transcriptome response of juvenile *Ancherythroculter nigrocauda* fed diets with different protein levels. *Comparative Biochemistry and Physiology Part D: Genomics and Proteomics.* 31: 100592.

Magnuson-Ford K, Ingram T, Redding DW, Mooers AØ (2009). Rockfish (Sebastes) that are evolutionarily isolated are also large, morphologically distinctive and vulnerable to overfishing. *Biological Conservation.* 142(8): 1787–1796.

Martin SA, Douglas A, Houlihan DF, Secombes CJ (2010). Starvation alters the liver transcriptome of the innate immune response in Atlantic salmon (*Salmo salar*). *BMC Genomics.* 11(1): 418.

Martin SA, Król E (2017). Nutrigenomics and immune function in fish: New insights from omics technologies. *Developmental & Comparative Immunology.* 75: 86–98.

Martínez H, Barrachina S, Castillo M, Tárraga J, Medina I, Dopazo J, Quintana-Ortíz ES (2018). A framework for genomic sequencing on clusters of multicore and manycore processors. *The International Journal of High Performance Computing Applications.* 32(3): 393–406.

Matsuda M, Nagahama Y, Shinomiya A, Sato T, Matsuda C, Kobayashi T, … Hori H (2002). DMY is a Y-specific DM-domain gene required for male development in the medaka fish. *Nature.* 417(6888): 559–563.

Mazurais D, Darias M, Zambonino-Infante JL, Cahu CL (2011). Transcriptomics for understanding marine fish larval development. *Canadian Journal of Zoology.* 89(7): 599–611.

Mekuchi M, Asakura T, Kikuchi J (2019). New aquaculture technology based on host-symbiotic co-metabolism. In: *Marine Metagenomics.* Springer, Singapore, 189–228.

Meyer A, Van de Peer Y (2005). From 2R to 3R: Evidence for a fish-specific genome duplication (FSGD). *Bioessays.* 27(9): 937–945.

Morais S, Silva T, Cordeiro O, Rodrigues P, Guy DR, Bron JE, … Tocher DR (2012). Effects of genotype and dietary fish oil replacement with vegetable oil on the intestinal transcriptome and proteome of Atlantic salmon (*Salmo salar*). *BMC Genomics.* 13(1): 448.

Naylor RL, Hardy RW, Bureau DP, Chiu A, Elliott M, Farrell AP, … Nichols PD (2009). Feeding aquaculture in an era of finite resources. *Proceedings of the National Academy of Sciences.* 106(36): 15103–15110.

Nielsen EE, Hemmer-Hansen JAKOB, Larsen PF, Bekkevold D (2009). Population genomics of marine fishes: Identifying adaptive variation in space and time. *Molecular Ecology.* 18(15): 3128–3150.

Pujolar JM, Marino IA, Milan M, Coppe A, Maes GE, Capoccioni F, … Cramb G (2012). Surviving in a toxic world: Transcriptomics and gene expression profiling in response to environmental pollution in the critically endangered European eel. *BMC Genomics.* 13(1): 507.

Qiao Q, Le Manach S, Huet H, Duvernois-Berthet E, Chaouch S, Duval C, … Lennon S (2016). An integrated omic analysis of hepatic alteration in medaka fish chronically exposed to cyanotoxins with possible mechanisms of reproductive toxicity. *Environmental Pollution.* 219: 119–131.

Reinartz J, Bruyns E, Lin J-Z, Burcham T, Brenner S, Bowen S, Kramer M, Woychik R (2002). Massively parallel signature sequencing (MPSS) as a tool for in-depth quantitative gene expression profiling in all organisms. *Briefings in Functional Genomics*. 1(1): 95–104.

Renaut S, Bernatchez L (2011). Transcriptome-wide signature of hybrid breakdown associated with intrinsic reproductive isolation in lake whitefish species pairs (*Coregonus* spp. Salmonidae). *Heredity*. 106(6): 1003–1011.

Ryder OA (2005). Conservation genomics: Applying whole genome studies to species conservation efforts. *Cytogenetic and Genome Research*. 108(1–3): 6–15.

Rodriguez-Barreto D, Rey O, Uren-Webster TM, Castaldo G, Consuegra S, Garcia de Leaniz C (2019). Transcriptomic response to aquaculture intensification in Nile tilapia. *Evolutionary Applications*. 12(9): 1757–1771.

Schad J, Voigt CC, Greiner S, Dechmann DK, Sommer S (2012). Independent evolution of functional MHC class II DRB genes in New World bat species. *Immunogenetics*. 64(7): 535–547.

Schadt EE, Linderman MD, Sorenson J, Lee L, Nolan GP (2010). Computational solutions to large-scale data management and analysis. *Nature Reviews Genetics*. 11(9): 647.

Sun F, Liu S, Gao X, Jiang Y, Perera D, Wang X, … Dunham R (2013). Male-biased genes in catfish as revealed by RNA-Seq analysis of the testis transcriptome. *PloS ONE*. 8(7): e68452.

Tacchi L, Bickerdike R, Douglas A, Secombes CJ, Martin SA (2011). Transcriptomic responses to functional feeds in Atlantic salmon (*Salmo salar*). *Fish & Shellfish Immunology*. 31(5): 704–715.

Takehana Y, Matsuda M, Myosho T, Suster ML, Kawakami K, Shin T, … Hamaguchi S (2014). Co-option of Sox3 as the male-determining factor on the Y chromosome in the fish *Oryzias dancena*. *Nature Communications*. 5(1): 1–10.

Tao W, Chen J, Tan D, Yang J, Sun L, Wei J, … Wang D (2018). Transcriptome display during tilapia sex determination and differentiation as revealed by RNA-Seq analysis. *BMC Genomics*. 19(1): 363.

Tong C, Zhang C, Zhang R, Zhao K (2015). Transcriptome profiling analysis of naked carp (*Gymnocypris przewalskii*) provides insights into the immune-related genes in highland fish. *Fish & Shellfish Immunology*. 46(2): 366–377.

Ulloa PE, Medrano JF, Feijoo CG (2014). Zebrafish as animal model for aquaculture nutrition research. *Frontiers in Genetics*. 5: 313.

Vera LM, Metochis C, Taylor JF, Clarkson M, Skjaerven KH, Migaud H, Tocher DR (2017). Early nutritional programming affects liver transcriptome in diploid and triploid Atlantic salmon, *Salmo salar*. *BMC Genomics*. 18(1): 886.

Vesterlund L, Jiao H, Unneberg P, Hovatta O, Kere J (2011). The zebrafish transcriptome during early development. *BMC Developmental Biology*. 11(1): 30.

Wadapurkar RM, Vyas R (2018). Computational analysis of next generation sequencing data and its applications in clinical oncology. *Informatics in Medicine Unlocked*. 11: 75–82.

Wang Z, Du J, Lam SH, Mathavan S, Matsudaira P, Gong Z (2010). Morphological and molecular evidence for functional organization along the rostrocaudal axis of the adult zebrafish intestine. *BMC Genomics*. 11(1): 392.

Wang Z, Gerstein M, Snyder M (2009). RNA-Seq: A revolutionary tool for transcriptomics. *Nature Reviews Genetics*. 10(1): 57.

Wang J, Liu Y, Jiang S, Li W, Gui L, Zhou T, ... Chen L (2019). Transcriptomic and epigenomic alterations of Nile tilapia gonads sexually reversed by high temperature. *Aquaculture*. 508: 167–177.

Weber GJ, Sepúlveda MS, Peterson SM, Lewis SS, Freeman JL (2013). Transcriptome alterations following developmental atrazine exposure in zebrafish are associated with disruption of neuroendocrine and reproductive system function, cell cycle, and carcinogenesis. *Toxicological Sciences*. 132 (2): 458–466.

Wolfe KH (2001). Yesterday's polyploids and the mystery of diploidization. *Nature Review Genetics*. 2(5): 333–341.

World Bank (2013). Fish to 2030: Prospects for fisheries and aquaculture. *In Agriculture and Environmental Services Discussion Paper*. 3.

Wu S, Ren Y, Peng C, Hao Y, Xiong F, Wang G, ... Angert ER (2015). Metatranscriptomic discovery of plant biomass-degrading capacity from grass carp intestinal microbiomes. *FEMS Microbiology Ecology*. 91(10): fiv107.

Xiao J, Zhong H, Liu Z, Yu F, Luo Y, Gan X, Zhou Y (2015). Transcriptome analysis revealed positive selection of immune-related genes in tilapia. *Fish & Shellfish Immunology*. 44(1): 60–65.

Xu W, Jin J, Han D, Liu H, Zhu X, Yang Y, Xie S (2019). Physiological and transcriptomic responses to fishmeal-based diet and rapeseed meal-based diet in two strains of gibel carp (*Carassius gibelio*). *Fish Physiology and Biochemistry*. 45(1): 267–286.

Yang Y, Han T, Xiao J, Li X, Wang J (2018). Transcriptome analysis reveals carbohydrate-mediated liver immune responses in *Epinephelus akaara*. *Scientific Reports*. 8(1): 639.

Yano A, Guyomard R, Nicol B, Jouanno E, Quillet E, Klopp C, ... Guiguen Y (2012). An immune-related gene evolved into the master sex-determining gene in rainbow trout, *Oncorhynchus mykiss*. *Current Biology*. 22(15): 1423–1428.

Yasuike M, Iwasaki Y, Nishiki I, Nakamura Y, Matsuura A, Yoshida K, ... Fujiwara A (2018). The yellowtail (*Seriola quinqueradiata*) genome and transcriptome atlas of the digestive tract. *DNA Research*. 25(5): 547–560.

Yokoi H, Kobayashi T, Tanaka M, Nagahama Y, Wakamatsu Y, Takeda H, ... Ozato K (2002). Sox9 in a teleost fish, medaka (*Oryzias latipes*): Evidence for diversified function of Sox9 in gonad differentiation. *Molecular Reproduction and Development: Incorporating Gamete Research*. 63(1): 5–16.

Żarski D, Nguyen T, Le Cam A, Montfort J, Dutto G, Vidal MO, ... Bobe J (2017). Transcriptomic profiling of egg quality in sea bass (*Dicentrarchus labrax*) sheds light on genes involved in ubiquitination and translation. *Marine Biotechnology*. 19(1): 102–115.

Zhang X, Shi J, Sun Y, Zhu Y, Zhang Z, Wang Y (2019). Transcriptome analysis provides insights into differentially expressed genes and long noncoding RNAs involved in sex-related differences in Amur sturgeon (*Acipenser schrenckii*). *Molecular Reproduction and Development*. 86(2): 132–144.

4

Transcriptomics Applied in Research of Non-Communicable Diseases

Ana Karen González-Palomo
Centro de Biociencias, Universidad Autónoma de San Luis Potosí, Mexico

Juan Carlos Fernández-Macias
Coordinación para la aplicación de la Ciencia y la Tecnología (CIACYT), Universidad Autónoma de San Luis Potosí, Mexico

Velia Verónica Rangel-Ramírez
Centro de Biociencias, Universidad Autónoma de San Luis Potosí, Mexico

Jorge Armando Jimenez-Avalos
Unidad Guadalajara, Centro de Investigación y Asistencia en Tecnología y Diseño del Estado de Jalisco (CIATEJ), Mexico

Juan Diego Cortés-Garcia
Facultad de Medicina, Universidad Autónoma de San Luis Potosí, Mexico

CONTENTS

4.1 Introduction

The most prevalent non-communicable diseases (NCDs) include different types of cancer and cardiovascular, respiratory and neurodegenerative

DOI: 10.1201/9781003212416-5

diseases. All of them have quite complex etiologies, where a great variety of factors are involved, of which genetic and environmental factors stand out. According to the World Health Organization (WHO), globally, 23% of all deaths could be prevented through healthier environments; also early life exposure to environmental risks, such as chemicals and air pollutants, might increase NCD risk through the life course of individuals (WHO, 2017). However, these factors do not act independently, and therefore the gene-environment interaction plays a key role in the development and study of these diseases. The former issue means that a genetic abnormality may be needed for a disease to occur, but the disease would not exist without the presence of an environmental risk factor (Willett, 2002).

Throughout the history of evolution, human beings have adapted to the different environments in which they live and, at the same time, react to the potential environmental hazards to which they are exposed at a certain point in life. These responses can be translated into activation or suppression of certain genes, alterations in messenger ribonucleic acid (mRNA) levels, epigenetic changes and post-translational modifications, among others (Olden & White, 2005). However, in transcripts from the first expression of the phenotype, and in all living beings, the interaction of the environment with the genome modulates the production of these transcripts, so this interaction is of crucial importance in the study of transcriptomics.

Transcriptomics allow us to determine the expression patterns of ribonucleic acid (RNA) transcripts across the genome, offering further fascinating clues to the functions of genetic variants and a platform to identify and quantify a set of genes in a given biological state (Yates et al., 2009). An interesting application of transcriptomics is the comparison of gene expression patterns in different biological states, such as disease and health, resulting in a signature of genes with altered expression under certain conditions. These altered transcripts can also be quantitatively assessed as new biomarkers (Pedrotty et al., 2012). To perform a transcriptomic study, transcriptomics technologies such as RNA sequencing (RNA-seq) are employed to capture almost the totality of the transcriptome (Wang, Gerstein, & Snyder 2009). RNA-seq can explore the complete content of a cellular transcriptome, allowing the detection of a wide range of RNA isoforms, as well as new RNA molecules; facilitating the handling and analysis of data, and identifying new disease markers.

This chapter describes some NCDs, their risk factors and the application of transcriptomics in the investigation of these diseases (Figure 4.1).

4.1.1 Non-Communicable Diseases

Non-communicable diseases (NCDs) are a global health problem that is increasing due to the aging of the population and current lifestyles, such as sedentary lifestyle and poor diet. The WHO estimates that NCDs accounted

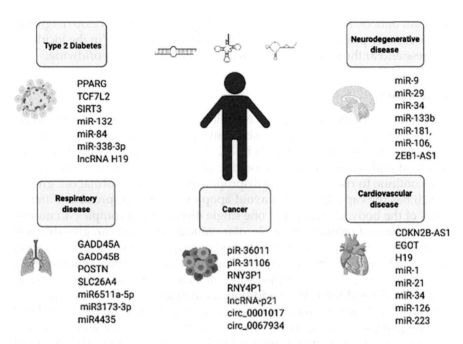

FIGURE 4.1

Transcriptomics applied in research of non-communicable disease. Model of RNA transcripts analyzed by the RNA-seq technique in cancer, type 2 diabetes, neurodegenerative, respiratory and cardiovascular diseases. PPARG (Peroxisomal Proliferative Activated Receptor Gamma), TCF7L2 (Transcription Factor 7-like 2), SIRT3 (Sirtuin 3), GADD45A and GADD45B (growth arrest and DNA gene), POSTN (Periostin), SLC26A4 (pendrin), miR (microRNA), RNY (long Ro-associated non-coding RNA), lncRNAs (long non-coding RNA), (circRNAs) circular RNA.

for 63% of all global deaths, 80% of which occur in low- and middle-income countries (WHO, 2019b). In addition, NCDs have generated a significant public economic expenditure for the therapeutic treatment of the population by health systems, generating a strong alarming socioeconomic impact (Simiao Chen et al., 2018). The more common NCDs are diabetes, cancer, cardiovascular, chronic respiratory, kidney and neurodegenerative diseases (CDC, 2018; WHO, 2019b). There are several NCD risk factors such as tobacco use, the harmful use of alcohol, poor nutrition, physical inactivity and environmental and occupational exposures (Arora et al., 2018; WHO, 2019b; Wu et al., 2015). These factors cause changes at the metabolic and physiological levels, resulting in overweight or obesity, hyperglycemia, hyperlipidemia and raised blood pressure, which can also contribute significantly to the development of NCDs (Arora et al., 2018). Because of this, the United Nations has established the Sustainable Development Goals (SDG) and within its purposes is the reduction of deaths caused by NCDs through prevention and treatment (United Nations, 2019). Therefore, providing relevant data on the discovery of new molecules helps to develop diagnostic procedures to prevent NCDs, which will lead to a reduction in the incidence of these and at the same time to increase the quality of life worldwide.

4.1.2 Cancer

According to the WHO, around 9.6 million people worldwide die from cancer in 2018, representing one in six deaths globally and becoming the second cause of death worldwide (WHO, 2019a). Those facts reflect a necessity to attend and address this situation due to the knowledge that this number will continue to rise. Cancer is a disproportionate and abnormal cell growth which has the capacity to both avoid apoptosis and also spread to different parts of the body. Cancer is not one single disease but a complex accumulation of different cellular and molecular pathogenesis that can give different clinical characteristics. The most-studied or suspected risk factors for cancer can include age, alcohol, cancer-causing substances, chronic inflammation, diet, hormones, immunosuppression, infectious agents, obesity, radiation, sunlight and tobacco. In addition, an increased risk of cancer has been observed with environmental exposure to certain substances considered carcinogenic, such as aflatoxins, arsenic, cadmium, benzene and pesticides, among others (International Agency for Research on Cancer, 2018; National Cancer Institute, 2018). Moreover, one of the principal features of cancer is its heterogeneity at inter- and intra-tumor levels; however, the heterogeneity among tumor-infiltrating cells and immune system cells in the microenvironment have key roles in tumor growth, angiogenesis, immune evasion, metastasis and responses to various therapies (X. Ren et al., 2018). Gold treatments for cancer are chemotherapy and targeted therapies; however, as we know, drug resistance is common for most tumors. Therefore, there is a need to develop new strategies that can help reach a personalized diagnosis. By means of understanding the composition tumor and the dynamics in the microenvironment, we can help to improve outcomes through early diagnosis, classification and monitorization of treatment response and disease progression and stop drug resistance.

The transcriptomic profile is a new tool that helps to provide more knowledge about cancer. This approach has the ability to detect whole gene expression levels and the diverse elements which are part of the RNA world. Several studies have demonstrated the application of transcriptomics to detect biomarkers of different types of cancer. Following are some examples of these biomarkers, in breast cancer (BC) it has been found that some RNAs called piRNAs (piR), piR-36011, piR-31106 and piR-36717 are differentially expressed in hormone-responsive BC, but also between cancer and normal breast tissue specimens (Hashim et al., 2015). Furthermore, piR-4987, piR-20365, piR-20485 and piR-20582 were confirmed to be up-regulated in BC using deep sequencing data and qPCR (quantitative Polymerase Chain

Reaction). Of these, piR-4987 was associated with positive lymph node status, making it a candidate as a biomarker in BC (G. Huang et al., 2013). Another study analyzed piR-1245, finding it overexpressed in colorectal cancer (CRC); patients with elevated piR-1245 levels were susceptible to metastasis and had a lower overall survival rate (Weng et al., 2018). piR-651 has been found to be highly up-regulated in gastric cancer (GC) tissues and in other different human cancer cell lines, including hepatic, cervical, breast, mesothelioma and lung (J. Cheng et al., 2011). Interestingly, of the piRNAs mentioned, piR-651 and piR-823 have been used as peripheral blood biomarkers for the detection of circulating cancer cells in GC patients, able to discern patients with GC from the healthy individuals (L. Cui et al., 2011). In gastric cancer, levels of piR-823 were up-regulated in serum and urine of renal carcinoma patients (n = 178) compared with healthy controls (n = 101), being the opposite in tumor tissue (Iliev et al., 2017). Another piRNA, piR-651 levels in serum in classical Hodgkin's lymphoma (cHL) patients were found to be lower than healthy controls and significantly elevated in the same patients after chemotherapy (Cordeiro et al., 2016).

On the other hand, another molecule of RNA that has been identified is YRNA (small non-coding RNA); its function is still poorly understood, although it appears to be necessary for DNA replication. Dhahbi et al. (2013) reported that YRNA accounts for 38% of cfRNA (cell-free circulating tumor RNA) in sera from BC patients. They found that 3´-end fragments of YRNAs were up-regulated, whereas 5´-end fragments were down-regulated in patients; those levels were associated with some clinicopathological characteristics, such as the epidermal growth factor receptor 2 (*HER2*), Estrogen receptor (*ER*) and progesterone receptor (*PR*) status, relapse and tumor stage. For another part, it was found that YRNAs accounted for more than 40% in plasma samples from melanoma patients. Three YRNA fragments (RNY3P1, RNY4P1 and RNY4P25) analyzed in 58 melanoma patients and 22 healthy controls demonstrated that these three fragments were up-regulated in pre-clinical melanoma (stage 0) compared to either healthy controls or more advanced-stage disease (Sole et al., 2019).

One of the most analyzed circulating transcripts, called long non-coding RNAs (lncRNAs), is the metastasis-associated lung adenocarcinoma transcript 1 (*MALAT1*). Fragments of *MALAT1* are found increased in plasma and serum samples from prostate cancer patients (S. Ren et al., 2013) and also in exosomes from the serum of patients with NSCLC (non-small-cell lung cancer) and in exosomes from the urine of patients with high-grade muscle-invasive urothelial bladder cancer (Berrondo et al., 2016; Zhang et al., 2017a) Another lncRNA, lncARSR, studied in renal cell carcinoma was shown to be overexpressed in patient plasma samples, being that those levels correlated with the presence of progressive disease (Qu et al., 2016). In hematological malignancies, lncRNA-p21 was found to be up-regulated in the plasma of Chronic Lymphocytic Leukemia (CLL) patients and *MALAT1*, HOX transcript antisense RNA (*HOTAIR*) and Growth Arrest Specific 5 (*GAS5*) in myeloma patients (Isin et al., 2014). Aside from being possible biomarkers, lncRNAs can also be used as therapeutic targets. One example is *SAMMSON* (Survival Associated Mitochondrial Melanoma-Specific Oncogenic Non-Coding RNA), which plays a central role in the growth of aggressive melanoma (Leucci et al., 2016). This lncRNA cannot be detected in normal melanocytes or other healthy tissues, so it could be a targeted therapeutic, hopefully without side effects. Not only does *SAMMSON* look like a good targeted therapeutic approach, but others like *LUNAR1* (Leukemia-Associated Non-Coding IGF1R Activator RNA 1), *NEAT1* (Nuclear Paraspeckle Assembly Transcript 1) and *PCGEM1* (Prostate-Specific Transcript 1) could also be identified as up- or down-regulated in cancer tissue, and are thus potential therapeutic targets (Arun et al., 2018).

Regarding circular RNA (circRNAs), gastric cancer is the most studied cancer type in the circRNA field, and many publications have described the down-regulation of circRNAs like circ_0000190 (Shijun Chen et al., 2017), circ_002059 (P. Li et al., 2015b) and circ_0001017, among others, all of them with the capability of being diagnostic biomarkers, and circ_0000745 (M. Huang et al., 2017) being suggested to have a prognostic approach. circ-LDLRAD3 was also up-regulated in plasma from 31 patients with pancreatic cancer and was associated with clinicopathological features of patients, such as lymphatic and venous invasion (Yang et al., 2017). circ_0000064 and circ_0013958 are related to tumor lymph node metastasis and TNM (tumor, node and metastasis) staging in lung cancer (Luo et al., 2017). The rise of circ_0067934 in esophageal cancer suggests the progression of tumor staging (Xia et al., 2016).

Another type of RNA are the small nuclear RNAs (snRNAs), and one the most studied is *RNU2-1f* (U2 small nuclear RNA fragments), which has been detected in the serum and plasma of lung, ovarian, pancreatic and colorectal cancers (Baraniskin et al., 2013; Köhler et al., 2016; Kuhlmann et al., 2014; Mazières et al., 2013). A study in lung cancer patients found that levels of *RNU2-1f* were associated with shorter median survival in stage III/IV patients (Köhler et al., 2016), and in another serum, samples taken from ovarian cancer patients showed higher levels of *RNU2-1f* compared to healthy individuals (Kuhlmann et al., 2014). Another snRNA that could be useful as a cancer biomarker is *U6*, which was investigated in addition to *SNORD44* (Small Nucleolar RNA, C/D Box 44) in the sera of 39 patients with breast cancer and sera from 40 healthy age-matched controls. The authors found that ratios of *U6* to *SNORD44* were higher in breast cancer patients, irrespective of ER status (Appaiah et al., 2011). As another example, Liao et al (2010) found a deregulation of *SNORA42* (small nucleolar RNAs) (among others) in plasma of non-small-cell lung cancer (NSCLC) patients; this RNA is believed to be implicated in tumorigenesis through two different pathways, a p53-dependent and a p53-independent (Liao et al., 2010). NSCLC patients with elevated *SNORA42* expression succumbed earlier from the disease (Miyamoto et al., 2015).

Although promising, RNA biomarkers still need more standardized operation procedures on many different levels which are analytical and clinical validation as well as an assessment of clinical utility (Shaw et al., 2015). As the ultimate goal of understanding the Omics is to help patients, we should not forget that there is a strong need for a multi-disciplinary team effort required by oncologists, genomics scientists, bioinformaticians, pathologists and genetic counselors so the information could be understood and translated in patient care. The success of cancer transcriptomics will be evaluated with the discovery of new drugs and molecular diagnostics, so new approaches in the Omics for the proteome and the metabolome can reveal additional knowledge of cancer biology making it more interesting to integrate all that data.

4.1.3 Type 2 Diabetes

Type 2 diabetes (T2D) is a complex endocrine disease that is characterized by loss and abnormal function of β cells in the pancreas, which cause insulin insufficiency (C. C. Thomas & Philipson, 2015) to overcome insulin resistance. Its etiology is associated with hyperglycemia, polydipsia, nephropathies and cardiovascular diseases (Saeedi Borujeni et al., 2018). In the world, more than 422 million people have diabetes, which caused 1.5 million deaths in 2016 (WHO, 2018). T2D is related to genetic and environmental factors, where β cell disruption is highlighted by the fact that participate in the biosynthesis and regulation of insulin. Also, the pancreas cells express many proteins related to glucose homeostasis and normal cell function. This is why the study of T2D pathology has been directed to the role of each pancreas cell line as α cells (glucagon), β cells (insulin), δ cells (somatostatin) and PP cells (pancreatic polypeptide).

In this sense, different reports have been describing the role of many differential gene expressions in T2D pathology as genes that participates in fasting glucose (FG) as the receptor for glucagon-like peptide 1 (*GLP1R*) that is associated with lower FG (Scott et al., 2016). Also, other genes as *G6PC2*, *GPSM1*, *SLC2A2*, *SLC30A8*, *RREB1* and *COBLL1* are related to FG and/or fasting insulin (FI) (Wessel et al., 2015). There are also genes that are implicated with body mass index (BMI) like *KCNJ11* which is negatively correlated with BMI; meanwhile, *TCF7L2* is not associated (Kirkpatrick et al., 2010). However, only a few genes are differentially expressed in T2D with respect to non-diabetic subjects.

RNA-seq technology is a useful tool that adjusts and discards multiple genes to find only the most relevant genes in the T2D status (Scott et al., 2016). Burton et al., compared the gene expression in subject with T2D and healthy subjects and found that three genes were the most relevant: *PPARG* (Peroxisomal Proliferative Activated Receptor Gamma), *KCNJ11* (Kir6.2 Component of the Pancreatic Beta-Cell KATP Channel) and *TCF7L2* (Transcription Factor 7-like 2), where *TCF7L2* has a stronger association with T2D through the Wnt-pathway that causes β cells dysfunction and so increases the diabetes risk. Also, *PPARG* and *KCNJ11* were associated with

T2D-susceptibility (Burton et al., 2007); however, other genes showed less association with T2D, thus evidencing the need to include the pancreas into the analysis. To determine the pancreas' role in the physiopathology of T2D, the tissue of patients was analyzed. In this sense, Fadista et al. (2014) detected 1619 genes related to HbA1c (glycated hemoglobin A1c) levels. Of these, the most expressed were *SLC30A8* (solute carrier family 30 (zinc transporter), member8), *G6PC2* (glucose-6-phosphatase catalytic 2) and *PCSK1* (proprotein convertase subtilisin/kexin type 1). Meanwhile, *RASGRP1* (RAS guanyl releasing protein 1) *RFX3* (transcription factor) and *NNT* (nicotinamide nucleotide transhydrogenase) were associated with lower HbA1c and higher insulin secretion, so it is believed that they may participate in regulating insulin secretion (Fadista et al., 2014).

In the same way, transcriptome analyses of isolated pancreas cells provide information of each cell population in T2D. Different reports showed the specific-single cell genes that differentiate the islets of the pancreas. The α cells, which synthesize glucagon, express genes GC (vitamin D binding protein), *ARX, GCG, DPP4* and *IRX2* (Ackermann et al., 2016), where noncoding GC is related to gestational diabetes (Y. Wang et al., 2015) and insulin sensitivity (Hirai et al., 2000). Other genes also expressed are *LOXL4, FAP* (Segerstolpe et al., 2016), *PLCE1, TMEM236, IGFBP2, COTL1, SPOCK3* and *ARRDC4* (Xin et al., 2016); these genes are involved in pathways that regulate the Ca^{2+} and cAMP cytoplasmic concentrations, and *GPR119* is associated with insulin and GLP-1 release that translate in a β cell mass increase in mice (Chu et al., 2008). The β cells, which synthesize insulin, differentially expressed genes as *MAFA, INS, IGF2, CHODL* and *SLC27A6* (Ackermann et al., 2016), where *CHODL* participates in the protein transporter (Zelensky & Gready, 2005) and INS is the insulin gene. Other genes expressed in these cells are *IAPP, DCYAP1, PDX1, NKX6-1, MEG3* (Segerstolpe et al., 2016), *RGS16* and *DLK1* (Xin et al., 2016); these genes participate in the metabolism of the glucose. The δ cells, which produce somatostatin, highest express *UNC5B, GABRB3, GABRG2, CASR, FFAR4/GPR120, KCNJ2* and *LEPR* genes (Segerstolpe et al., 2016), as well as *SST, BCHE, HHEX* and *RPL7P19* (Xin et al., 2016); these genes are related to cell proliferation.

Interestingly, β cells have a poor antioxidant system, so these cells are very sensitive to oxidative stress, which could be related to T2D physiopathology. Thus, the hyperglycemic condition is involved in increased Reactive Oxygen Species (ROS) in different tissues; in the pancreas, ROS may be the cause of β cells dysfunction by decreased INS gene expression and its transcription factor *PDX1* (Poitout et al., 2006). Also, the *PPARGC1A* gene promotes ROS-increased levels (Besseiche et al., 2018), which in consequence induce a down-regulated *SIRT3* expression and an increase of *CASP3* levels, these events translate into a toxic effect by chronic glucose exposure (Y. Zhou et al., 2017). Therefore, different markers have been used to determine the oxidative status in T2D patients, such as the Advanced Glycation End Products (*AGEs*) that are augmented and associated with T2D complications (Nowotny et al., 2015). Other markers could explain better the effect of hyperglycemic condition on β cells such as miRNAs or lncRNAs.

With respect to miRNAs, small non-coding RNAs of 20 –21 nucleotides in length, are posttranscriptional regulators and are implicated in the development and function of the pancreas (LaPierre & Stoffel, 2017). In this sense, Saeedi Borujeni et al. (2018) describe that they can be divided into two groups in T2D: some miRNA that correlate with β cells dysfunction previously T2D such as miR-132, miR-84 and miR-338-3p, and a second group expressed in β cells of T2D patients such as miR-34a, miR-146a, miR-199a-3p, miR-203, miR-210 and miR-383 (Saeedi Borujeni et al., 2019). For another part, in energy homeostasis, miR-375 is a regulator of α and β cell functions and participates in the inhibition of insulin secretion (Poy et al., 2009). Meanwhile, miR-7a controls the proliferation of β cells (Latreille et al., 2014), and miR-9 decreases the insulin secretion (Hu et al., 2018). Also, miR-21, miR-34a and miR-146 participate in the pro-inflammatory state associated with T2D (Roggli et al., 2010).

With respect to lncRNA, they are non-coding RNAs greater than 200 nucleotides in length whose function is to participate in the different cellular process as a post-transcriptional regulator, protein localization, cell cycle and others (Carninci et al., 2005). In T2D, the lncRNA *H19* and *MALAT1* is associated with gestational diabetes due to its involvement in deficient structure and function of pancreas (Ding et al., 2012; Y. Zhang, 2018). In the same way, Morán et al. (2012) describe the importance of the lncRNA HI-LNC25 in the function of β cells when diminishing the transcription factor *GLIS3* (Morán et al., 2012). Other lncRNAs involved in T2D are *LOC283177*, related with secretion and synthesis of insulin (Fadista et al., 2014), *LIN01099* with expression restricted mainly to β cells (Segerstolpe et al., 2016) and *MEG3* associated with T2D complications (Yan et al., 2014).

4.1.4 Neurodegenerative Diseases

Neurodegenerative diseases affect millions of people worldwide. These are a heterogeneous group of diseases that affect the nervous system and are characterized by a progressive neuronal loss in specific areas of the brain or anatomical systems. Neurodegeneration can be found in many different levels of neuronal circuitry ranging from molecular to systemic (W.-W. Chen et al., 2016; Dugger & Dickson, 2017).

Many of these diseases have an important genetic factor, and the greatest risk is aging, but sometimes the origin is a medical condition such as alcoholism, a tumor or stroke. Other causes may include dietary and lifestyle factors, toxins, chemicals and viruses. Also, in a great number of cases, the cause is unknown (W.-W. Chen et al., 2016; JPND, 2017).

Neurotoxic substances are those capable of causing adverse effects to the central nervous system, the peripheral nervous system, and the sense organs.

Although most of the effects are reversible (nausea, dizziness, behavior alterations) some of them cause an alteration in neuronal function, progressive degeneration and death of neurons, leading to the development of neurodegenerative diseases (JPND, 2017; Musgrove et al., 2015). The most common of neurodegenerative diseases in which environmental factors are associated are Alzheimer's disease, Parkinson's disease and amyotrophic lateral sclerosis (Brown et al., 2005; Cannon & Greenamyre, 2011).

One in nine people over 65 suffer from Alzheimer's disease (AD). This is the most common form of dementia, affecting cognition, memory, language and behavior. This disease is characterized by amyloid-β (Aβ) deposition and neurofibrillary tangle (NFT) formation. Even though the highest percentage of risk to AD is attributable to aging and genetic factors (the most studied risk gene is one form of the apolipoprotein E, APOE ϵ4), a considerable part for its etiology is due to environmental factors (De-Paula et al., 2012; G. Li et al., 2015a; Yegambaram et al., 2015).

Although the epidemiological associations between exposures to environmental pollutants and AD are not clearly studied, evidence suggests that induction of NFT through phosphorylation of tau protein, the generation of reactive oxygen species and the stimulation of Aβ deposition could be triggered by exposure to toxic metals (aluminum, copper), organochlorine and organophosphate insecticides, polybrominated diphenyl ethers (PBDEs) and particulate matter (Al-Mousa & Michelangeli, 2012; Bradner et al., 2013; Cacciottolo et al., 2017; Gauthier et al., 2001; Liu et al., 2006; Moulton & Yang, 2012; Parrón et al., 2011; Shcherbatykh & Carpenter, 2007).

The increasing recognition of the complexity of the interactions in the regulation of gene expression with transcription factors, the coding and non-coding RNAs and alternative splicing lead to proposals for a better understanding of the molecular pathogenesis of AD (Twine et al., 2011).

As mentioned previously, the main genetic factor is APOE ϵ4, and the mechanisms with respect to AD involve neurotoxicity, Aβ aggregation, tau phosphorylation, synaptic plasticity and neuroinflammation (Qian et al., 2017; Wadhwani et al., 2019). Besides APOE ϵ4, around 24 mutations in amyloid precursor protein (*APP*), 185 in presenilin 1 (*PSEN1*) and 13 in presenilin 2 (*PSEN2*) have been reported to be pathogenic for AD (Cruts et al., 2012; Humphries & Kohli, 2014; Ringman & Coppola, 2013). RNA-seq analysis of the parietal cortex shows a total of 4073 isoforms up-regulated and 558 down-regulated and reveals alternatively spliced isoforms related to lipid metabolism (Mills et al., 2013).

Several studies demonstrated that genes related to inflammation have increased expression in AD across several brain regions (Avramopoulos et al., 2011; Paouri & Georgopoulos, 2019; Parachikova et al., 2007). In the same way, 5 of the 20 AD GWAS-identified genes are involved in inflammation: Complement receptor type 1 (*CR1*), Sialic Acid-Binding Ig-Like Lectin 3 (*CD33*), Major Histocompatibility Complex, Class II, DR Beta 5-DR Beta

1 (*HLA-DRB5-DRB1*), Ectonucleotide Pyrophosphatase/Phosphodiesterase Family Member 5 (*ENPP5*) and Myocyte Enhancer Factor 2C (*MEF2C*) (Lambert et al., 2013).

A study of RNA-seq with differential expression analysis identified four down-regulated miRNAs (miR-184, miR-34c-3p, miR-375 and miR-132-5p) in AD patients with respect to control hippocampal samples, and microarray miRNA profiling confirmed the down-regulation of the miR-132/212 family consistent with a role in AD pathogenesis (Annese et al., 2018). Aberrant regulation of miRNA-dependent gene expression has been implicated in the molecular events responsible for Aβ production, NFT formation and neurodegeneration. It has been suggested that miRNAs such as miR-9, miR-107, miR-29, miR-34, miR-181, miR-106, miR-146a, miR132 and miR153 could regulate the expression of APP direct and indirect, the alternative splicing of APP mRNA and modulate tau phosphorylation (Gupta et al., 2017; Prendecki & Dorszewska, 2014; Viswambharan et al., 2017).

The transcriptomic profiles of patients with AD include susceptibility genes linked to the pathways of APP and protein tau, immune response, inflammation, cell migration, lipid transport, endocytosis, hippocampal synaptic function, cytoskeletal function, axonal transport, regulation of gene expression and post-translational modification of proteins, and microglial and myeloid cell function.

Furthermore, Parkinson's disease (PD) is the most prevalent neurodegenerative movement disorder characterized by extensive degeneration of dopaminergic neurons in the nigrostriatal system. Neurochemical and neuropathological analyses clearly indicate that oxidative stress, mitochondrial dysfunction, neuroinflammation and impairment of the ubiquitin-proteasome system (UPS) are major mechanisms of dopaminergic degeneration (Kanthasamy et al., 2010; Moretto & Colosio, 2011).

So far, no determinant factor has been found for its triggering, although some studies have identified characteristics that increase a person's risk of developing PD, including gender, age, race, and genetic factors (Pringsheim et al., 2014). However, neurotoxic exposure that prompts neuronal apoptosis is considered a leading risk factor in the development of this disease for over 90% of sporadic PD cases (Di Monte, Lavasani, & Manning-Bog 2002; Gerlach & Riederer 1999; Kanthasamy et al., 2010; Moretto & Colosio 2011; Thomas 2009). Several studies have shown the importance of exposure to pesticides and other toxics with the development of PD. Neurotoxins such as MPP+ (Kaul et al., 2003), dieldrin (Kitazawa et al., 2004), manganese (Latchoumycandane, 2005), rotenone (Sherer et al., 2003; Testa et al., 2005) and paraquat (McCormack et al., 2005), activate major events of the apoptotic pathway, including cytochrome C release, caspase-3 activation and DNA fragmentation.

Oligonucleotide microarrays are the most widely used transcriptomics studies in PD (Borrageiro et al., 2018). These assays have shown a differentiated expression of genes between PD cases and controls. In 2018 a microarray analysis found 359 up-regulated and 460 down-regulated differentially expressed genes (DEGs) in 41 PD patients, in the same study a protein-protein interaction (PPI) network analysis showed the top differentially expressed genes with higher degrees: Estrogen Receptor 1 (*ESR1*), Mechanistic Target of Rapamycin Kinase (*MTOR*), Ataxia Telangiectasia Mutated (*ATM*), Tumor Necrosis Factor Receptor Superfamily Member 5 (*CD40*), Thymidine Kinase 2 (*TK2*), Mitogen-Activated Protein Kinase 14 (*MAPK14*), Phosphatase and Tensin Homolog (*PTEN*), Intercellular Adhesion Molecule 1 (*ICAM1*), Aurora Kinase A (*AURKA*) and Catalytic Subunit of Protein Kinase DNA- Activated (*PRKDC*) (Tan et al., 2018).

RNA-seq and miRNA arrays or small RNA sequencing approaches are other common techniques for the transcriptomic study of PD (Borrageiro et al., 2018). Numerous studies have reported abnormal expression of miRNAs in PD patients comparing with healthy controls: miR-133b, miR-34b, miR-34c, mir-153, miR-7 and miR-433. These miRNAs related with levels of α-Synuclein (the primary component of Lewy bodies) and survival and differentiation of dopaminergic neurons (Zhang et al., 2017b; Leggio et al., 2017; Martinez and Peplow 2017; Mouradian 2012).

The most relevant pathways which were consistently found to be altered in PD include dopamine metabolism, mitochondrial function, oxidative stress, protein degradation, neuroinflammation, vesicular transport and synaptic transmission.

Another progressive and fatal disease of the nervous system that affects neurons responsible for controlling muscle movement is amyotrophic lateral sclerosis (ALS). ALS belongs to a wider group of disorders known as motor neuron diseases. In ALS, motor neurons degenerate or die and lose communication with muscles, which begin to weaken progressively and atrophy; this ends up affecting the ability to move, speak, eat and breathe (Bento-Abreu et al., 2010; NINDS, 2014). Ten percent of ALS cases are inherited, associated with dominant mutations in or deletion of the superoxide dismutase 1 gene (SOD1). In contrast, the majority of ALS cases (90 percent or more) are sporadic. This is an example of why it is considered that gene-environmental interaction provides a plausible explanation and plays an important role in the development of this disease. Exposure to lead, mercury, pesticides and electromagnetic fields have all been cited as potential risk factors for ALS (Johnson & Atchison, 2009; Kamel et al., 2012; Martínez-Sámano et al., 2012; Trojsi et al., 2013; Weisskopf et al., 2009). These environmental factors that have been related to ALS have in common that they may induce a cellular increase of reactive oxygen species with consequent DNA damage and apoptosis.

In microarray analysis, a gene group associated with cytoskeletal and mitochondrial dysfunction in the motor cortex of patients with ALS have been identified (Lederer et al., 2007). In another study, a long non-coding RNA (lncRNA) characterization in ALS patients was realized, 293 differentially expressed (DE) lncRNA were identified and, furthermore, three lncRNAs reported in the top 10 are described as antisense of transcription-related genes: *ZEB1-AS1* is antisense of *ZEB1* transcription factor, *ZBTB11-AS1* is the antisense of *ZBTB11* gene, involved in the DNA binding and in transcriptional regulation, and XXbac-BPG252P9.10 is the antisense of *IER3*, transcription factor of the nuclear factor-kappa-B/rel (NF-kappa-B) family. In the same study, RNA-seq data showed 87 differentially expressed mRNAs compared to healthy subjects, 30 of which were down-regulated while 57 were up-regulated, related to apoptotic process and transcription regulation, mainly to NF-kappa-B (Gagliardi et al., 2018).

On the other hand, miRNA microarrays show that among the miRNAs that were found consistently deregulated are miR-206, miR-9, miR-133, miR-143, miR-218, miR-146, miR-149, miR-335 and miR-338 (Dardiotis et al., 2018; Ferrante & Conti, 2017; Hamzeiy et al., 2018; Hoye et al., 2017; Krokidis & Vlamos, 2018; Ricci et al., 2018; Viswambharan et al., 2017). Several studies have proposed miRNA-206 as a potential circulating biomarker candidate for ALS (de Andrade et al., 2016; Toivonen et al., 2014; Waller et al., 2017; Williams et al., 2009).

In order to clarify the pathophysiological pathways contributing to motor neuron degeneration, diverse molecular mechanisms have been proposed, including oxidative stress, glutamate excitotoxicity, mitochondrial dysfunction, dysregulation of RNA processing, protein aggregation, disordered axonal and inflammation.

4.1.5 Chronic Respiratory Diseases

Chronic respiratory diseases (CRDs) are diseases that affect the lungs and airways. The most common CRDs are asthma, chronic obstructive pulmonary disease (COPD), occupational lung diseases and pulmonary hypertension (Centers for Disease Control and Prevention, 2017; Obeidat et al., 2019; World Health Organization, 2019). According to WHO, every year an estimated 3.9 million deaths are due to CRD, chronic obstructive pulmonary disease (COPD) being the main cause due to the fact that it is the third leading cause of death worldwide (World Health Organization, 2017b).

Chronic obstructive pulmonary disease (COPD) is the cause of 5% of deaths worldwide, this disease is characterized by a persistent reduction of airflow (Steiling et al., 2013; World Health Organization, 2019), and its risk factors are tobacco smoke (including passive exposure) (Duffy & Criner, 2019; Forey et al., 2011), environmental exposure such as indoor air pollution and outdoor air pollution (H.-C. Huang et al., 2019; Pathak et al., 2019; Zhao et al., 2019), and occupational chemicals exposure (Boschetto et al., 2006; Syamlal et al., 2019). The diagnosis of COPD is easy and fast through spirometry, simply by the presence of persistent airflow limitation, as determined by the forced expiratory volume in 1s (FEV1) and the ratio of FEV1 to forced vital capacity (FVC) (Andreeva et al., 2017; Faner et al., 2019). However, COPD is a complex disease at the clinical and molecular level because there are phenotypic and pathologic changes that can vary across individuals. Currently, transcriptome analysis is a tool used for the development of clinically relevant biomarkers and the identification of molecular subtypes, in order to improve therapies and customize the management of the disease (Obeidat et al., 2019; Steiling et al., 2013; Wadhwa et al., 2019).

The gene expression profile associated with the severity of COPD has been analyzed using the standard technique of microarrays in patients with critical respiratory disease compared to patients with non-critical exacerbations. It was found in the blood of these patients that the severity of the disease is related to a greater expression of serine protease neutrophil genes (NSPs) like Elastase, neutrophil expressed (*ELANE*), cathepsin G (*CTSG*) and proteinase 3 (*PRTN3*), which are involved in antimicrobial defense but also regulate tissue damage in lungs, and as a consequence, can explain respiratory failure in these patients (Almansa et al., 2012). Similarly, it has been determined the gene expression profile using the microarray technique in samples of sputum and circulating blood in patients with COPD to know if it can be replaced by the pulmonary parenchyma sample, to identify non-invasive biomarkers. However, no relationship was found between the transcriptome of lung tissue in patients with COPD and the assays done in blood and sputum (Faner et al., 2019).

Recently, important studies have used RNA-seq to identify the difference of gene expression between patients with COPD and control subjects. In this regard, 2312 genes and splicing variants were identified and related to different processes such as reduction of oxidative phosphorylation, modulation of the catabolism of proteins and deregulation of transcription. This has helped to increase the understanding of the pathophysiology of the disease (Kim et al., 2015). Also, it has been found that 17 up-regulated microRNAs have interactions with 18 down-regulated mRNAs, such as miR6511a-5p–NT5E, miR3173-3p–SDK1, miR4435–TNS1 and miR641–PCDH7, in primary normal human bronchial epithelial (NHBE) and COPD bronchial epithelial (DHBE) cells, this by RNA-seq, this newly found interactions might play important roles in the bronchial epithelium of COPD (W. A. Chang et al., 2018). Other relevant genes of COPD pathogenesis have been identified through the study of the transcriptome. Both *GADD45A* and *GADD45B* (growth arrest and DNA gene) genes are involved in DNA damage repair and tolerance genes (DDRT), and is believed that they are relevant to COPD pathogenesis as they are implicated in cell cycle arrest, apoptosis and cellular senescence (Sauler et al., 2018).

Likewise, asthma is a chronic lung disease that, like COPD, affects millions of people around the world. It is characterized by chronic inflammation, respiratory tract remodeling, hyperreactivity, shortness of breath, wheezing, cough and limitation of expiratory airflow (Bush, 2019). The severity of the disease can be measured with a chronic reduction in FEV1 and peak expiratory flow (PEF) (Sawyer et al., 1998). Several risk factors have been associated with the development of asthma, including family history, exposure to allergens and air pollutants, smoking and obesity (American Lung Association, 2019; Toskala & Kennedy, 2015).

Next-generation sequencing (RNA-seq) has given a detailed characterization of the transcriptomic profile in asthma. One study showed the comparison of the transcriptomic profile of bronchial biopsies, finding 46 genes that are differentially expressed between asthma and controls. These include B-cell CLL/lymphoma 2 (*BCL2*), Periostin (*POSTN*) and pendrin (*SLC26A4*). Interestingly, pendrin was one of the most up-regulated genes and has been a link to the pathophysiology of asthma. (Yick et al., 2013a).

On the other hand, three main therapies are used to treat patients with asthma, including short-acting and long-acting β2-agonists, corticosteroids, and leukotriene agonists (Gibeon & Menzies-Gow, 2013; Morjaria et al., 2018). However, these drugs are not effective for all patients. Transcriptomic studies by RNA-seq have shown the efficacy of some drugs for asthma treatment. In patients who were treated for 14 days with prednisolone, it was identified the gene cysteine-rich secretory protein LCCL domain-containing, 2 (*CRISPLD2*) in endobronchial biopsies after treatment. *CRISPLD2* participates as a regulator of airway smooth muscle (ASM) cells because it encodes a protein involved in lung development. This work has revealed more about the mechanism of action of corticosteroids. It was found that Single Nucleotide Polymorphism (SNPs) in the *CRISPLD2* gene are involved with resistance to the effect of inhaled corticosteroids and the bronchodilator response in patients with asthma (Himes et al., 2014). Similarly, using the same conditions with corticosteroids in patients with asthma, other genes like the Family with Sequence Similarity 129 Member A (*FAM129A*) and Synaptopodin 2 (*SYNPO2*) were associated with an improvement in the effect on airways (Yick et al., 2013b). For another part, the mechanism of action of paenoiflorin, the major active component of the Chinese peony (*Paeonia lactiflora*), has been described by transcriptomic analysis in asthmatic BALB/c mice, showing that paeoniflorin has a beneficial effect on fatty acid metabolism, inflammatory response, oxidative stress and local adhesion on model *in vivo* (Shou et al., 2019).

Although there is a variety of evidence in which transcriptomics have been applied in chronic respiratory diseases, there is still a field to be explored to know about gene regulations and other RNA transcripts in these diseases.

4.1.6 Cardiovascular Diseases

Despite many notable advances in prevention, treatment and management, cardiovascular diseases (CVDs) remain the leading cause of death around the globe. CVDs are NCDs that are responsible for 17.9 million deaths every year, representing 31% of all global deaths; they occur mainly in low- and middle-income countries, and its prevalence is on the rise (World Health Organization, 2017a). According to the WHO, ischemic heart disease and stroke were ranked first and second in the list of "Top 10 Causes of Death" in 2016 (World Health Organization, 2018). The rate of mortality and morbidity of CVDs worldwide have serious multi-level economic consequences in terms of disability and healthcare costs (Gheorghe et al., 2018).

CVDs are a group of disorders that lead to an integral loss in function of the heart, cardiovascular structures, and the circulatory system; they include ischemic heart disease, heart failure and cerebrovascular disease (World Health Organization, 2017a). Numerous risk factors have been associated with increased risk of CVDs, the major ones are those linked with habits and lifestyle, including high blood pressure, high blood cholesterol, high body mass index and smoking (Benjamin et al., 2019). In this regard, the American Heart Association (AHA) launched a program focused on tracking seven key health factors and behaviors, called "Life's Simple 7," as a strategic plan to achieve the Impact Goal: "By 2020, to improve the cardiovascular health of all Americans by 20% while reducing deaths from cardiovascular diseases and stroke by 20%" (Mozaffarian et al., 2015). The Life's Simple 7 behaviors are no smoking, physical activity, healthy diet, body weight and control of cholesterol, blood pressure and blood sugar (Benjamin et al., 2019), as the main determinants of CVDs. However, CVDs are multifactorial disorders, and many other factors play a potential role in their incidence, prevalence and severity, such as environmental stressors. In 2004, the AHA stated that exposure to particulate matter (PM) air pollution contributes to cardiovascular morbidity and mortality (Brook et al., 2010). Furthermore, worldwide evidence suggested that 29% of the risk of stroke is attributable to air pollution (Benjamin et al., 2019). In addition to behavioral and environmental factors, CVDs have a complex genetic basis related to perturbations in the expression of numerous genes, transcription factors or contractile proteins (Marian et al., 2016). Recent advances in the technologies and informatics to process large biological data sets (omics data) have opened the possibility to understand diseases at a molecular level (Manzoni et al., 2018).

In the heart, certain heterogenous cell types contribute to the development and progression of pathological changes in CVDs (Ackers-Johnson et al., 2018). Since its conception 10 years ago, through RNA-seq using a single mouse blastomere, Tang et al. (2009) could detect the expression of the entire transcriptome of a single cell. Similarly, Gladka et al. (2018) performed an RNA-seq from both healthy and diseased human cardiac tissue. They could identify multiple subpopulations within a certain cell type and found differential gene expression of cytoskeleton-associated protein 4 as a novel marker for activated fibroblasts that positively correlates with myofibroblast markers in humans, suggesting a key role in fibroblast activation during injured heart (Gladka et al., 2018). This technology has been applied to understanding and tracking the pathogeneses of other CVDs. Disorders like heart failure are preceded by a process of hypertrophy caused by persistent overload (Frey & Olson, 2003). During this process, cardiomyocytes activate signaling pathways that lead to failing phenotype characterized by elongation and contractile force reduction (Haque & Wang, 2017). Studies have demonstrated that cellular phenotypes determine cardiac tissue function, and phenotypes are highly dependent on transcriptomic profiles. Nomura et al. (2018) report an integrated transcriptomic, morphological and epigenetic analysis of the p53/Metf2/Nrf2 signaling axis at the transition to heart failure.

The lncRNAs have attracted increased attention in the progression and development of CVDs. Greco et al., identified and characterized lncRNAs from left ventricle biopsies of 18 patients with ischemic heart failure who were compared with controls. They found out that 14 lncRNAs were deregulated in non-end-stage heart failure patients, whereas 9 of these lncRNAs (*CDKN2B-AS1/ANRIL, EGOT, H19, HOTAIR, LOC285194/TUSC7, RMRP, RNY5, SOX2-OT* and *SRA1*) were found in end-stage failing hearts (Greco et al., 2016). Using RNA-seq technology, Di Salvo et al. (2015) compared lncRNA expression profile from 22 explanted human hearts and identified 84,793 total messenger RNAs; of these, 2,085 were lncRNAs. In comparison with donor hearts, 3.7% of lncRNAs were differentially expressed (Di Salvo et al., 2015).

Circulating microRNAs (miRNAs) have been recognized as key biomarkers of heart failure (HF), due to their involvement in the development and progression of HF. In patients with HF, deregulated levels of 30 circulating miRNAs have described (S. S. Zhou et al., 2018). A marked miR-21 upregulation and a down-regulation of miR-1 were observed in patients with symptomatic HF. Correlation between miR-21 and galectin-3, a novel biomarker of adverse cardiac remodeling, confirms its role in HF progression, whereas miR-1 expression correlates with the severity of HF (Sygitowicz et al., 2015). Increasing evidence has been shown involving the key role of miR-210, briefly miR-1/-21/-23/-126/-423-5p, in a cohort where 338 patients with HF (n = 236 acute HF; n = 58 non-acute HF; and n = 44 stable HF) were analyzed. Authors found an association of low circulating miR-423-5p with poor long-term outcome in AHF patients, identifying miR-423-5p expression as a prognostic biomarker (Seronde et al., 2015). In addition, in two independent cohorts of 2203 participants in different continents, miR-1254 and miR-1306-5p were associated with higher risk of death and HF hospitalization (Bayés-Genis et al., 2018). Changes in profile expression of several miRNAs have been reported in acute myocardial infarction (AMI), including down-regulation of miR-106, miR-197 and miR-223, as well as up-regulation of miR-1, miR-133, miR-21, miR-29b, miR-192, miR-194, miR-34a, miR-208, miR-499, miR-423, miR-126, miR-134, miR-328 and miR-486 (S. S. Zhou et al., 2018). In a similar way to miR-1, miR-133 is a master regulator of cardiac muscle development. It has been reported that in AMI patients, miR-133 is transiently up-regulated in a 4.4-fold increase, and returned to basal levels after one week of AMI in comparison with control individuals (R. Wang et al., 2011; Widera et al., 2011).

Several studies have associated gene expression profiling with blood pressure traits such as hypertension. In a meta-analysis of six independent studies of global gene expression from whole blood of 7017 individuals that have not received anti-hypertensive treatment, they found 30 for genes differentially expressed. These include the transcription factor *FOS* and Prostaglandin-Endoperoxide Synthase 2 (*PTGS2*), which were previously reported implicated in hypertension and other novel genes involved in CVDs such as Granzyme B (*GZMB*), Annexin A1 (*ANXA1*), Transmembrane protein 43 (*TMEM43*), Potassium Voltage-Gated Channel Subfamily J Member 2 (*KCNJ2*) and Myeloid cell leukemia-1(*MCL1*). These results could explain 5%–9% of interindividual variability in blood pressure traits (Huan et al., 2015).

4.1.7 Future Perspectives and Conclusions

As the magnitude of NCDs continue to grow worldwide, the need for accurate technology to measure and track the progression of NCDs is urgently demanded. For these reasons, the recent efforts to identify biomarkers with diagnostic and prognostic implications are necessary to prevent and treat NCDs in an affordable manner. The "omics" era to integrate data coming from transcriptomic analysis could participate in determining biomarkers related to early prognosis, tracking and staging of disease and even give insights about the treatment at a personalized level.

References

Ackermann, A. M., Wang, Z., Schug, J., Naji, A., & Kaestner, K. H. (2016). Integration of ATAC-seq and RNA-seq identifies human alpha cell and beta cell signature genes. *Molecular Metabolism*, 5(3), 233–244. doi:10.1016/j.molmet.2016.01.002

Ackers-Johnson, M., Tan, W. L. W., & Foo, R. S. Y. (2018). Following hearts, one cell at a time: Recent applications of single-cell RNA sequencing to the understanding of heart disease. *Nature Communications*, 9(1), 8–11. doi:10.1038/s41467-018-06894-8

Al-Mousa, F., & Michelangeli, F. (2012). Some commonly used brominated flame retardants cause Ca2+-ATPase inhibition, beta-amyloid peptide release and apoptosis in SH-SY5Y neuronal cells. *PLoS ONE*, 7(4), e33059. doi:10.1371/journal.pone.0033059

Almansa, R., Socias, L., Sanchez-Garcia, M., Martín-Loeches, I., Del Olmo, M., Andaluz-Ojeda, D., Bobillo, F., Rico, L., Herrero, A., Roig, V., San-Jose, C. A., Rosich, S., Barbado, J., Disdier, C., De Lejarazu, R. O., Gallegos, M. C., Fernandez, V., & Bermejo-Martin, J. F. (2012). Critical COPD respiratory illness is linked to increased transcriptomic activity of neutrophil proteases genes. *BMC Research Notes, 5*(Imi). doi:10.1186/1756-0500-5-401

American Lung Association. (2019). *Asthma Risk Factors.* https://www.lung.org/lung-health-and-diseases/lung-disease-lookup/asthma/asthma-symptoms-causes-risk-factors/asthma-risk-factors.html

Andreeva, E., Pokhaznikova, M., Lebedev, A., Moiseeva, I., Kuznetsova, O., & Degryse, J. M. (2017). Spirometry is not enough to diagnose COPD in epidemiological studies: A follow-up study. *Npj Primary Care Respiratory Medicine, 27*(1). doi:10.1038/s41533-017-0062-6

Annese, A., Manzari, C., Lionetti, C., Picardi, E., Horner, D. S., Chiara, M., Caratozzolo, M. F., Tullo, A., Fosso, B., Pesole, G., & D'Erchia, A. M. (2018). Whole transcriptome profiling of late-onset Alzheimer's disease patients provides insights into the molecular changes involved in the disease. *Scientific Reports, 8*(1), 4282. doi:10.1038/s41598-018-22701-2

Appaiah, H. N., Goswami, C. P., Mina, L. A., Badve, S., Sledge, G. W., Liu, Y., & Nakshatri, H. (2011). Persistent upregulation of U6:SNORD44 small RNA ratio in the serum of breast cancer patients. *Breast Cancer Research, 13*(5), R86. doi:10.1186/bcr2943

Arora, M., Mathur, C., Rawal, T., Bassi, S., Lakshmy, R., Nazar, G. P., Gupta, V. K., Park, M. H., & Kinra, S. (2018). Socioeconomic differences in prevalence of biochemical, physiological, and metabolic risk factors for non-communicable diseases among urban youth in Delhi, India. *Preventive Medicine Reports, 12*(47), 33–39. doi:10.1016/j.pmedr.2018.08.006

Arun, G., Diermeier, S. D., & Spector, D. L. (2018). Therapeutic targeting of long non-coding RNAs in cancer. *Trends in Molecular Medicine, 24*(3), 257–277. doi:10.1016/j.molmed.2018.01.001

Avramopoulos, D., Szymanski, M., Wang, R., & Bassett, S. (2011). Gene expression reveals overlap between normal aging and Alzheimer's disease genes. *Neurobiology of Aging, 32*(12), 2319.e27–2319.e34. doi:10.1016/j.neurobiolaging.2010.04.019

Baraniskin, A., Nöpel-Dünnebacke, S., Ahrens, M., Jensen, S. G., Zöllner, H., Maghnouj, A., Wos, A., Mayerle, J., Munding, J., Kost, D., Reinacher-Schick, A., Liffers, S., Schroers, R., Chromik, A. M., Meyer, H. E., Uhl, W., Klein-Scory, S., Weiss, F. U., Stephan, C., & Hahn, S. A. (2013). Circulating U2 small nuclear RNA fragments as a novel diagnostic biomarker for pancreatic and colorectal adenocarcinoma. *International Journal of Cancer, 132*(2), 48–58. doi:10.1002/ijc.27791

Bayés-Genis, A., Lanfear, D., de Ronde, M., Lupón, J., Leenders, J., Liu, Z., Zuithoff, N., Eijkemans, M., Zamora, E., De Antonio, M., Zwinderman, A., Pinto-Sietsma, S., & Pinto, Y. (2018). Prognostic value of circulating microRNAs on heart failure-related morbidity and mortality in two large diverse cohorts of general heart failure patients. *pdf. European Journal of Heart Failure, 20*(1), 67–75.

Benjamin, E. J., Muntner, P., Alonso, A., Bittencourt, M. S., Callaway, C. W., Carson, A. P., Chamberlain, A. M., Chang, A. R., Cheng, S., Das, S. R., Delling, F. N., Djousse, L., Elkind, M. S. V., Ferguson, J. F., Fornage, M., Jordan, L. C., Khan, S. S., Kissela, B. M., Knutson, K. L., & Virani, S. S. (2019). Heart disease and stroke statistics-2019 update: A report from the American heart association. *Circulation, 139*(10), e56–e528. doi:10.1161/CIR.0000000000000659

Bento-Abreu, A., Van Damme, P., Van Den Bosch, L., & Robberecht, W. (2010). The neurobiology of amyotrophic lateral sclerosis. *European Journal of Neuroscience, 31*(12), 2247–2265. doi:10.1111/j.1460-9568.2010.07260.x

Berrondo, C., Flax, J., Kucherov, V., Siebert, A., Osinski, T., Rosenberg, A., Fucile, C., Richheimer, S., & Beckham, C. J. (2016). Expression of the long non-coding RNA HOTAIR correlates with disease progression in bladder cancer and is contained in bladder cancer patient urinary exosomes. *PLoS ONE, 11*(1), 1–21. doi:10.1371/journal.pone.0147236

Besseiche, A., Riveline, J. P., Delavallée, L., Foufelle, F., Gautier, J. F., & Blondeau, B. (2018). Oxidative and energetic stresses mediate beta-cell dysfunction induced by PGC-1α. *Diabetes and Metabolism, 44*(1), 45–54. doi:10.1016/j.diabet.2017.01.007

Borrageiro, G., Haylett, W., Seedat, S., Kuivaniemi, H., & Bardien, S. (2018). A review of genome-wide transcriptomics studies in Parkinson's disease. *European Journal of Neuroscience, 47*(1), 1–16. doi:10.1111/ejn.13760

Boschetto, P., Quintavalle, S., Miotto, D., Lo Cascio, N., Zeni, E., & Mapp, C. E. (2006). Chronic obstructive pulmonary disease (COPD) and occupational exposures. *Journal of Occupational Medicine and Toxicology (London, England), 1*(1), 11. doi:10.1186/1745-6673-1-11

Bradner, J. M., Suragh, T. A., Wilson, W. W., Lazo, C. R., Stout, K. A., Kim, H. M., Wang, M. Z., Walker, D. I., Pennell, K. D., Richardson, J. R., Miller, G. W., & Caudle, W. M. (2013). Exposure to the polybrominated diphenyl ether mixture DE-71 damages the nigrostriatal dopamine system: Role of dopamine handling in neurotoxicity. *Experimental Neurology, 241*, 138–147. doi:10.1016/j.expneurol.2012.12.013

Brook, R. D., Rajagopalan, S., Pope, C. A., Brook, J. R., Bhatnagar, A., Diez-Roux, A. V., Holguin, F., Hong, Y., Luepker, R. V., Mittleman, M. A., Peters, A., Siscovick, D., Smith, S. C., Whitsel, L., & Kaufman, J. D. (2010). Particulate matter air pollution and cardiovascular disease. *Circulation, 121*(21), 2331–2378. doi:10.1161/CIR.0b013e3181dbece1

Brown, R. C., Lockwood, A. H., & Sonawane, B. R. (2005). Neurodegenerative diseases: An overview of environmental risk factors. *Environmental Health Perspectives, 113*(9), 1250–1256. doi:10.1289/ehp.7567

Burton, P. R., Clayton, D. G., Cardon, L. R., Craddock, N., Deloukas, P., Duncanson, A., Kwiatkowski, D. P., McCarthy, M. I., Ouwehand, W. H., Samani, N. J., Todd, J. A., Donnelly, P., Barrett, J. C., Davison, D., Easton, D., Evans, D., Leung, H. T., Marchini, J. L., Morris, A. P., & Compston, A. (2007). Genome-wide association study of 14,000 cases of seven common diseases and 3,000 shared controls. *Nature, 447*(7145), 661–678. doi:10.1038/nature05911

Bush, A. (2019). Pathophysiological mechanisms of asthma. *Frontiers in Pediatrics, 7*(March), 1–17. doi:10.3389/fped.2019.00068

Cacciottolo, M., Wang, X., Driscoll, I., Woodward, N., Saffari, A., Reyes, J., Serre, M. L., Vizuete, W., Sioutas, C., Morgan, T. E., Gatz, M., Chui, H. C., Shumaker, S. A., Resnick, S. M., Espeland, M. A., Finch, C. E., & Chen, J. C. (2017). Particulate air pollutants, APOE alleles and their contributions to cognitive impairment in older women and to amyloidogenesis in experimental models. *Translational Psychiatry, 7*(1), e1022. doi:10.1038/tp.2016.280

Cannon, J. R., & Greenamyre, J. T. (2011). The role of environmental exposures in neurodegeneration and neurodegenerative diseases. *Toxicological Sciences, 124*(2), 225–250. doi:10.1093/toxsci/kfr239

Carninci, P., Kasukawa, T., Katayama, S., Gough, J., Frith, M. C., Maeda, N., & Zhu, S. (2005). The transcriptional landscape of the mammalian genome. *Science, 309*(5740), 1559–1563.

Centers for Disease Control and Prevention (CDC). (2017). *Chronic Respiratory Disease.* https://www.cdc.gov/niosh/respiratory/health.html

Centers for Disease Control and Prevention (CDC). (2018). *Global Noncommunicable Diseases (NCDs).* https://www.cdc.gov/globalhealth/healthprotection/ncd/about.html

Chang, W. A., Tsai, M. J., Jian, S. F., Sheu, C. C., & Kuo, P. L. (2018). Systematic analysis of transcriptomic profiles of COPD airway epithelium using next-generation sequencing and bioinformatics. *International Journal of COPD, 13*, 2387–2398. doi:10.2147/COPD.S173206

Chen, Shijun, Li, T., Zhao, Q., Xiao, B., & Guo, J. (2017). Using circular RNA hsa_circ_0000190 as a new biomarker in the diagnosis of gastric cancer. *Clinica Chimica Acta, 466*, 167–171. doi:10.1016/j.cca.2017.01.025

Chen, Simiao, Kuhn, M., Prettner, K., & Bloom, D. E. (2018). The macroeconomic burden of noncommunicable diseases in the United States: Estimates and projections. *PLoS ONE, 13*(11), 1–14. doi:10.1371/journal.pone.0206702

Chen, W.-W., Zhang, X., & Huang, W.-J. (2016). Role of neuroinflammation in neurodegenerative diseases (Review). *Molecular Medicine Reports, 13*(4), 3391–3396. doi:10.3892/mmr.2016.4948

Cheng, J., Guo, J. M., Xiao, B. X., Miao, Y., Jiang, Z., Zhou, H., & Li, Q. N. (2011). PiRNA, the new non-coding RNA, is aberrantly expressed in human cancer cells. *Clinica Chimica Acta, 412*(17–18), 1621–1625. doi:10.1016/j.cca.2011.05.015

Chu, Z. L., Carroll, C., Alfonso, J., Gutierrez, V., He, H., Lucman, A., Pedraza, M., Mondala, H., Gao, H., Bagnol, D., Chen, R., Jones, R. M., Behan, D. P., & Leonard, J. (2008). A role for intestinal endocrine cell-expressed G protein-coupled receptor 119 in glycemic control by enhancing glucagon-like peptide-1 and glucose-dependent insulinotropic peptide release. *Endocrinology, 149*(5), 2038–2047. doi:10.1210/en.2007-0966

Cordeiro, A., Navarro, A., Gaya, A., Díaz-Beyá, M., Farré, B. G., Castellano, J. J., Fuster, D., Martínez, C., Martínez, A., & Monzó, M. (2016). PiwiRNA-651 as marker of treatment response and survival in classical Hodgkin lymphoma. *Oncotarget, 7*(29). doi:10.18632/oncotarget.10015

Cruts, M., Theuns, J., & Van Broeckhoven, C. (2012). Locus-specific mutation databases for neurodegenerative brain diseases. *Human Mutation, 33*(9), 1340–1344. doi:10.1002/humu.22117

Cui, L., Lou, Y., Zhang, X., Zhou, H., Deng, H., Song, H., Yu, X., Xiao, B., Wang, W., & Guo, J. (2011). Detection of circulating tumor cells in peripheral blood from patients with gastric cancer using piRNAs as markers. *Clinical Biochemistry, 44*(13), 1050–1057. doi:10.1016/j.clinbiochem.2011.06.004

Dardiotis, E., Aloizou, A.-M., Siokas, V., Patrinos, G. P., Deretzi, G., Mitsias, P., Aschner, M., & Tsatsakis, A. (2018). The role of microRNAs in patients with amyotrophic lateral sclerosis. *Journal of Molecular Neuroscience, 66*(4), 617–628. doi:10.1007/s12031-018-1204-1

De-Paula, V. J., Radanovic, M., Diniz, B. S., & Forlenza, O. V. (2012). *Alzheimer's disease. Sub-Cellular Biochemistry, 65*, 329–352. doi:10.1007/978-94-007-5416-4_14

de Andrade, H. M. T., de Albuquerque, M., Avansini, S. H., de Rocha S. C., Dogini, D. B., Nucci, A., Carvalho, B., Lopes-Cendes, I., & França, M. C. (2016). MicroRNAs-424 and 206 are potential prognostic markers in spinal onset amyotrophic lateral sclerosis. *Journal of the Neurological Sciences, 368*, 19–24. doi:10.1016/j.jns.2016.06.046

Dhahbi, J. M., Spindler, S. R., Atamna, H., Boffelli, D., Mote, P., Martin, D. I. (2013). 5′-YRNA fragments derived by processing of transcripts from specific YRNA genes and pseudogenes are abundant in human serum and plasma. *Physiological Genomics, 45*, 990–998. doi:10.1152/physiolgenomics.00129.2013

Di Monte, D. A., Lavasani, M., & Manning-Bog, A. B. (2002). Environmental factors in Parkinson's disease. *NeuroToxicology, 23*(4–5), 487–502. doi:10.1016/S0161-813X(02)00099-2

Di Salvo, T. G., Guo, Y., Su, Y. R., Clark, T., Brittain, E., Absi, T., Maltais, S., & Hemnes, A. (2015). Right ventricular long noncoding RNA expression in human heart failure. *Pulmonary Circulation, 5*(1), 135–161. doi:10.1086/679721

Ding, G. L., Wang, F. F., Shu, J., Tian, S., Jiang, Y., Zhang, D., Wang, N., Luo, Q., Zhang, Y., Jin, F., Leung, P. C. K., Sheng, J. Z., & Huang, H. F. (2012). Transgenerational glucose intolerance with Igf2/H19 epigenetic alterations in mouse islet induced by intrauterine hyperglycemia. *Diabetes, 61*(5), 1133–1142. doi:10.2337/db11-1314

Duffy, S. P., & Criner, G. J. (2019). Chronic obstructive pulmonary disease: Evaluation and management. *Medical Clinics of North America, 103*(3), 453–461. doi:10.1016/j.mcna.2018.12.005

Dugger, B. N., & Dickson, D. W. (2017). Pathology of neurodegenerative diseases. *Cold Spring Harbor Perspectives in Biology, 9*(7), a028035. doi:10.1101/cshperspect.a028035

Fadista, J., Vikman, P., Laakso, E. O., Mollet, I. G., Esguerra, J. L., Taneera, J., Storm, P., Osmark, P., Ladenvall, C., Prasad, R. B., Hansson, K. B., Finotello, F., Uvebrant, K., Ofori, J. K., Di Camillo, B., Krus, U., Cilio, C. M., Hansson, O., Eliasson, L., & Groop, L. (2014). Global genomic and transcriptomic analysis of human pancreatic islets reveals novel genes influencing glucose metabolism. *Proceedings of the National Academy of Sciences, 111*(38), 13924–13929. doi:10.1073/pnas.1402665111

Faner, R., Morrow, J. D., Casas-Recasens, S., Cloonan, S. M., Noell, G., López-Giraldo, A., Tal-Singer, R., Miller, B. E., Silverman, E. K., Agustí, A., & Hersh, C. P. (2019). Do sputum or circulating blood samples reflect the pulmonary transcriptomic differences of COPD patients? A multi-tissue transcriptomic network META-analysis. *Respiratory Research, 20*(1), 1–11. doi:10.1186/s12931-018-0965-y

Ferrante, M., & Conti, G. O. (2017). Environment and neurodegenerative diseases: An update on miRNA role. *MicroRNA, 6*(3). doi:10.2174/2211536606666170811151503

Forey, B. A., Thornton, A. J., & Lee, P. N. (2011). Systematic review with meta-analysis of the epidemiological evidence relating smoking to COPD, chronic bronchitis and emphysema. *BMC Pulmonary Medicine, 11*(1), 36. doi:10.1186/1471-2466-11-36

Frey, N., & Olson, E. N. (2003). Cardiac hypertrophy: The good, the bad, and the ugly. *Annual Review of Physiology, 65*(1), 45–79. doi:10.1146/annurev.physiol.65.092101.142243

Gagliardi, S., Zucca, S., Pandini, C., Diamanti, L., Bordoni, M., Sproviero, D., Arigoni, M., Olivero, M., Pansarasa, O., Ceroni, M., Calogero, R., & Cereda, C. (2018). Long non-coding and coding RNAs characterization in peripheral blood mononuclear cells and spinal cord from amyotrophic lateral sclerosis patients. *Scientific Reports, 8*(1), 2378. doi:10.1038/s41598-018-20679-5

Gauthier, E., Fortier, I., Courchesne, F., Pepin, P., Mortimer, J., & Gauvreau, D. (2001). Environmental pesticide exposure as a risk factor for Alzheimer's disease: A case-control study. *Environmental Research, 86*(1), 37–45. doi:10.1006/enrs.2001.4254

Gerlach, M., & Riederer, P. F. (1999). Time sequences of dopaminergic cell death in Parkinson's disease: Indications for neuroprotective studies. *Advances in Neurology, 80*, 219–225.

Gheorghe, A., Griffiths, U., Murphy, A., Legido-Quigley, H., Lamptey, P., & Perel, P. (2018). The economic burden of cardiovascular disease and hypertension in low- and middle-income countries: A systematic review. *BMC Public Health, 18*(1), 1–11. doi:10.1186/s12889-018-5806-x

Gibeon, D., & Menzies-Gow, A. (2013). Recent changes in the drug treatment of allergic asthma. *Clinical Medicine, Journal of the Royal College of Physicians of London, 13*(5), 477–481. doi:10.7861/clinmedicine.13-5-477

Gladka, M. M., Molenaar, B., De Ruiter, H., Versteeg, D., Tsui, H., Lacraz, G., Van Der Elst, S., Huibers, M., Van Oudenaarden, A., & Van Rooij, E. (2018). 239Single-cell sequencing of the healthy and diseased heart reveals Ckap4 as a new modulator of fibroblasts activation. *Cardiovascular Research, 114*(suppl_1), S61. doi:10.1093/cvr/cvy060.167

Greco, S., Zaccagnini, G., Perfetti, A., Fuschi, P., Valaperta, R., Voellenkle, C., Castelvecchio, S., Gaetano, C., Finato, N., Beltrami, A. P., Menicanti, L., & Martelli, F. (2016). Long noncoding RNA dysregulation in ischemic heart failure. *Journal of Translational Medicine, 14*(1), 1–14. doi:10.1186/s12967-016-0926-5

Gupta, P., Bhattacharjee, S., Sharma, A. R., Sharma, G., Lee, S.-S., & Chakraborty, C. (2017). miRNAs in Alzheimer disease - A therapeutic perspective. *Current Alzheimer Research, 14*(11). doi:10.2174/1567205014666170829101016

Hamzeiy, H., Suluyayla, R., Brinkrolf, C., Janowski, S. J., Hofestädt, R., & Allmer, J. (2018). Visualization and analysis of miRNAs implicated in amyotrophic lateral sclerosis within gene regulatory pathways. *Studies in Health Technology and Informatics, 253*, 183–187. doi:10.3233/978-1-61499-896-9-183

Haque, Z. K., & Wang, D. Z. (2017). How cardiomyocytes sense pathophysiological stresses for cardiac remodeling. *Cellular and Molecular Life Sciences, 74*(6), 983–1000. doi:10.1007/s00018-016-2373-0

Hashim, A., Rizzo, F., Marchese, G., Ravo, M., Tarallo, R., Nassa, G., Giurato, G., Santamaria, G., Cordella, A., Cantarella, C., & Weisz, A. (2015). RNA sequencing identifies specific PIWI-interacting small non-coding RNA expression patterns in breast cancer. *Oncotarget, 5*(20). doi:10.18632/oncotarget.2476

Himes, B. E., Jiang, X., Wagner, P., Hu, R., Wang, Q., Klanderman, B., Whitaker, R. M., Duan, Q., Lasky-Su, J., Nikolos, C., Jester, W., Johnson, M., Panettieri, R. A., Tantisira, K. G., Weiss, S. T., & Lu, Q. (2014). RNA-Seq transcriptome profiling identifies CRISPLD2 as a glucocorticoid responsive gene that modulates cytokine function in airway smooth muscle cells. *PLoS ONE, 9*(6). doi:10.1371/journal.pone.0099625

Hirai, M., Suzuki, S., Hinokio, Y., Hirai, A. K. I., Chiba, M., Akai, H., Suzuki, C., & Toyota, T. (2000). Variations in vitamin D-binding protein (Group-specific component protein) are associated with fasting plasma tolerance. *Journal of Clinical Endocrinology & Metabolism, 85*(5), 1951–1953.

Hoye, M. L., Koval, E. D., Wegener, A. J., Hyman, T. S., Yang, C., O'Brien, D. R., Miller, R. L., Cole, T., Schoch, K. M., Shen, T., Kunikata, T., Richard, J.-P., Gutmann, D. H., Maragakis, N. J., Kordasiewicz, H. B., Dougherty, J. D., & Miller, T. M. (2017). MicroRNA profiling reveals marker of motor neuron disease in ALS models. *The Journal of Neuroscience, 37*(22), 5574–5586. doi:10.1523/JNEUROSCI.3582-16.2017

Hu, D., Wang, Y., Zhang, H., & Kong, D. (2018). Identification of miR-9 as a negative factor of insulin secretion from beta cells. *Journal of Physiology and Biochemistry, 74*(2), 291–299. doi:10.1007/s13105-018-0615-3

Huan, T., Esko, T., Peters, M. J., Pilling, L. C., Schramm, K., Schurmann, C., Chen, B. H., Liu, C., Joehanes, R., Johnson, A. D., Yao, C., Ying, S. xia, Courchesne, P., Milani, L., Raghavachari, N., Wang, R., Liu, P., Reinmaa, E., Dehghan, A., & van Duijn, C. M. (2015). A meta-analysis of gene expression signatures of blood pressure and hypertension. *PLoS Genetics, 11*(3), 1–29. doi:10.1371/journal. pgen.1005035

Huang, G., Hu, H., Xue, X., Shen, S., Gao, E., Guo, G., Shen, X., & Zhang, X. (2013). Altered expression of piRNAs and their relation with clinicopathologic features of breast cancer. *Clinical and Translational Oncology, 15*(7), 563–568. doi:10.1007/ s12094-012-0966-0

Huang, H.-C., Lin, F. C.-F., Wu, M.-F., Nfor, O. N., Hsu, S.-Y., Lung, C.-C., Ho, C.-C., Chen, C.-Y., & Liaw, Y.-P. (2019). Association between chronic obstructive pulmonary disease and PM2.5 in Taiwanese nonsmokers. *International Journal of Hygiene and Environmental Health*, March, 0–1. doi:10.1016/j.ijheh.2019.03.009

Huang, M., He, Y. R., Liang, L. C., Huang, Q., & Zhu, Z. Q. (2017). Circular RNA hsa-circ-0000745 may serve as a diagnostic marker for gastric cancer. *World Journal of Gastroenterology, 23*(34), 6330–6338. doi:10.3748/wjg.v23.i34.6330

Humphries, C., & Kohli, M. A. (2014). Rare variants and transcriptomics in Alzheimer disease. *Current Genetic Medicine Reports, 2*(2), 75–84. doi:10.1007/ s40142-014-0035-9

Iliev, R., Fedorko, Mi., Machackova, T., Mlcochova, H., Svoboda, M., Pacik, D., Dolezel, J., Stanik, M., & Slaby, O. (2017). Expression levels of PIWI-interacting RNA, piR-823, are deregulated in tumor tissue, blood serum and urine of patients with renal cell carcinoma. *Anticancer Research, 36*(12), 6419–6424. doi:10.21873/ anticanres.11239

International Agency for Research on Cancer. (2018). *IARC Monographs on the Identification of Carcinogenic Hazards to Humans.* Occupational Exposures in Insecticide Application, and Some Pesticides. https://monographs.iarc.fr/ iarc-monographs-on-the-evaluation-of-carcinogenic-risks-to-humans-68/

Isin, M., Ozgur, E., Cetin, G., Erten, N., Aktan, M., Gezer, U., & Dalay, N. (2014). Investigation of circulating lncRNAs in B-cell neoplasms. *Clinica Chimica Acta, 431*, 255–259. doi:10.1016/j.cca.2014.02.010

Johnson, F. O., & Atchison, W. D. (2009). The role of environmental mercury, lead and pesticide exposure in development of amyotrophic lateral sclerosis. *NeuroToxicology, 30*(5), 761–765. doi:10.1016/j.neuro.2009.07.010

JPND. (2017). *What Is Neurodegenerative Disease?*

Kamel, F., Umbach, D. M., Bedlack, R. S., Richards, M., Watson, M., Alavanja, M. C. R., Blair, A., Hoppin, J. A., Schmidt, S., & Sandler, D. P. (2012). Pesticide exposure and amyotrophic lateral sclerosis. *NeuroToxicology, 33*(3), 457–462. doi:10.1016/j. neuro.2012.04.001

Kanthasamy, A., Jin, H., Mehrotra, S., Mishra, R., Kanthasamy, A., & Rana, A. (2010). Novel cell death signaling pathways in neurotoxicity models of dopaminergic degeneration: Relevance to oxidative stress and neuroinflammation in Parkinson's disease. *NeuroToxicology, 31*(5), 555–561. doi:10.1016/j. neuro.2009.12.003

Kaul, S., Kanthasamy, A., Kitazawa, M., Anantharam, V., & Kanthasamy, A. G. (2003). Caspase-3 dependent proteolytic activation of protein kinase Cdelta mediates and regulates 1-methyl-4-phenylpyridinium (MPP+)-induced apoptotic cell death in dopaminergic cells: Relevance to oxidative stress in dopaminergic degeneration. *European Journal of Neuroscience, 18*(6), 1387–1401. doi:10.1046/j.1460-9568.2003.02864.x

Kim, W. J., Lim, J. H., Lee, J. S., Lee, S.-D., Kim, J. H., & Oh, Y.-M. (2015). Comprehensive analysis of transcriptome sequencing data in the lung tissues of COPD subjects. *International Journal of Genomics, 2015*, 1–9. doi:10.1155/2015/206937

Kirkpatrick, C. L., Marchetti, P., Purrello, F., Piro, S., Bugliani, M., Bosco, D., de Koning, E. J. P., Engelse, M. A., Kerr-Conte, J., Pattou, F., & Wollheim, C. B. (2010). Type 2 diabetes susceptibility gene expression in normal or diabetic sorted human alpha and beta cells: Correlations with age or BMI of islet donors. *PLoS ONE, 5*(6). doi:10.1371/journal.pone.0011053

Kitazawa, M., Anantharam, V., Kanthasamy, A., & Kanthasamy, A. G. (2004). Dieldrin promotes proteolytic cleavage of poly(ADP-ribose) polymerase and apoptosis in dopaminergic cells: Protective effect of mitochondrial anti-apoptotic protein Bcl-2. *NeuroToxicology, 25*(4), 589–598. doi:10.1016/j.neuro.2003.09.014

Köhler, J., Schuler, M., Gauler, T. C., Nöpel-Dünnebacke, S., Ahrens, M., Hoffmann, A. C., Kasper, S., Nensa, F., Gomez, B., Hahnemann, M., Breitenbuecher, F., Cheufou, D., Özkan, F., Darwiche, K., Hoiczyk, M., Reis, H., Welter, S., Eberhardt, W. E. E., Eisenacher, M., & Baraniskin, A. (2016). Circulating U2 small nuclear RNA fragments as a diagnostic and prognostic biomarker in lung cancer patients. *Journal of Cancer Research and Clinical Oncology, 142*(4), 795–805. doi:10.1007/s00432-015-2095-y

Krokidis, M. G., & Vlamos, P. (2018). Transcriptomics in amyotrophic lateral sclerosis. *Frontiers in Bioscience (Elite Edition), 10*, 103–121.

Kuhlmann, J. D., Baraniskin, A., Hahn, S. A., Mosel, F., Bredemeier, M., Wimberger, P., Kimmig, R., & Kasimir-Bauer, S. (2014). Circulating U2 small nuclear RNA fragments as a novel diagnostic tool for patients with epithelial ovarian cancer. *Clinical Chemistry, 60*(1), 206–213. doi:10.1373/clinchem.2013.213066

Lambert, J.-C., Ibrahim-Verbaas, C. A., Harold, D., Naj, A. C., Sims, R., Bellenguez, C., Jun, G., DeStefano, A. L., Bis, J. C., Beecham, G. W., Grenier-Boley, B., Russo, G., Thornton-Wells, T. A., Jones, N., Smith, A. V., Chouraki, V., Thomas, C., Ikram, M. A., Zelenika, D., & Amouyel, P. (2013). Meta-analysis of 74,046 individuals identifies 11 new susceptibility loci for Alzheimer's disease. *Nature Genetics, 45*(12), 1452–1458. doi:10.1038/ng.2802

LaPierre, M. P., & Stoffel, M. (2017). MicroRNAs as stress regulators in pancreatic beta cells and diabetes. *Molecular Metabolism, 6*(9), 1010–1023. doi:10.1016/j.molmet.2017.06.020

Latchoumycandane, C. (2005). Protein kinase C is a key downstream mediator of manganese-induced apoptosis in dopaminergic neuronal cells. *Journal of Pharmacology and Experimental Therapeutics, 313*(1), 46–55. doi:10.1124/jpet.104.078469

Latreille, M., Hausser, J., Stützer, I., Zhang, Q., Hastoy, B., Gargani, S., Kerr-Conte, J., Pattou, F., Zavolan, M., Esguerra, J. L. S., Eliasson, L., Rülicke, T., Rorsman, P., & Stoffel, M. (2014). MicroRNA-7a regulates pancreatic β cell function. *Journal of Clinical Investigation, 124*(6), 2722–2735. doi:10.1172/JCI73066

Lederer, C. W., Torrisi, A., Pantelidou, M., Santama, N., & Cavallaro, S. (2007). Pathways and genes differentially expressed in the motor cortex of patients with sporadic amyotrophic lateral sclerosis. *BMC Genomics, 8*(1), 26. doi:10.1186/1471-2164-8-26

Leggio, L., Vivarelli, S., L'Episcopo, F., Tirolo, C., Caniglia, S., Testa, N., Marchetti, B., & Iraci, N. (2017). microRNAs in Parkinson's disease: From pathogenesis to novel diagnostic and therapeutic approaches. *International Journal of Molecular Sciences, 18*(12), 2698. doi:10.3390/ijms18122698

Leucci, E., Vendramin, R., Spinazzi, M., Laurette, P., Fiers, M., Wouters, J., Radaelli, E., Eyckerman, S., Leonelli, C., Vanderheyden, K., Rogiers, A., Hermans, E., Baatsen, P., Aerts, S., Amant, F., Van Aelst, S., Van Den Oord, J., De Strooper, B., Davidson, I., & Marine, J. C. (2016). Melanoma addiction to the long non-coding RNA SAMMSON. *Nature, 531*(7595), 518–522. doi:10.1038/nature17161

Li, G., Kim, C., Kim, J., Yoon, H., Zhou, H., & Kim, J. (2015a). Common pesticide, dichlorodiphenyltrichloroethane (DDT), increases amyloid-β levels by impairing the function of ABCA1 and IDE: Implication for Alzheimer's disease. *Journal of Alzheimer's Disease, 46*(1), 109–122. doi:10.3233/JAD-150024

Li, P., Chen, S., Chen, H., Mo, X., Li, T., Shao, Y., Xiao, B., & Guo, J. (2015b). Using circular RNA as a novel type of biomarker in the screening of gastric cancer. *Clinica Chimica Acta, 444*, 132–136. doi:10.1016/j.cca.2015.02.018

Liao, J., Yu, L., Mei, Y., Guarnera, M., Shen, J., Li, R., Liu, Z., & Jiang, F. (2010). Small nucleolar RNA signatures as biomarkers for non-small-cell lung cancer. *Molecular Cancer, 9*, 1–10. doi:10.1186/1476-4598-9-198

Liu, G., Huang, W., Moir, R. D., Vanderburg, C. R., Lai, B., Peng, Z., Tanzi, R. E., Rogers, J. T., & Huang, X. (2006). Metal exposure and Alzheimer's pathogenesis. *Journal of Structural Biology, 155*(1), 45–51. doi:10.1016/j.jsb.2005.12.011

Luo, Y. H., Zhu, X. Z., Huang, K. W., Zhang, Q., Fan, Y. X., Yan, P. W., & Wen, J. (2017). Emerging roles of circular RNA hsa_circ_0000064 in the proliferation and metastasis of lung cancer. *Biomedicine and Pharmacotherapy, 96*(November), 892–898. doi:10.1016/j.biopha.2017.12.015

Manzoni, C., Kia, D. A., Vandrovcova, J., Hardy, J., Wood, N. W., Lewis, P. A., & Ferrari, R. (2018). Genome, transcriptome and proteome: The rise of omics data and their integration in biomedical sciences. *Briefings in Bioinformatics, 19*(2), 286–302. doi:10.1093/BIB/BBW114

Marian, A. J., van Rooij, E., & Roberts, R. (2016). Genetics and genomics of single-gene cardiovascular diseases. *Journal of the American College of Cardiology, 68*(25), 2831–2849. doi:10.1016/j.jacc.2016.09.968

Martínez-Sámano, J., Torres-Durán, P. V., Juárez-Oropeza, M. A., & Verdugo-Díaz, L. (2012). Effect of acute extremely low frequency electromagnetic field exposure on the antioxidant status and lipid levels in rat brain. *Archives of Medical Research, 43*(3), 183–189. doi:10.1016/j.arcmed.2012.04.003

Martinez, B., & Peplow, P. (2017). MicroRNAs in Parkinson's disease and emerging therapeutic targets. *Neural Regeneration Research, 12*(12), 1945. doi:10.4103/1673-5374.221147

Mazières, J., Catherinne, C., Delfour, O., Gouin, S., Rouquette, I., Delisle, M. B., Prévot, G., Escamilla, R., Didier, A., Persing, D. H., Bates, M., & Michot, B. (2013). Alternative processing of the U2 small nuclear RNA produces a 19-22nt fragment with relevance for the detection of non-small cell lung cancer in human serum. *PLoS ONE, 8*(3). doi:10.1371/journal.pone.0060134

McCormack, A. L., Atienza, J. G., Johnston, L. C., Andersen, J. K., Vu, S., & Di Monte, D. A. (2005). Role of oxidative stress in paraquat-induced dopaminergic cell degeneration. *Journal of Neurochemistry, 93*(4), 1030–1037. doi:10.1111/j.1471-4159.2005.03088.x

Mills, J. D., Nalpathamkalam, T., Jacobs, H. I. L., Janitz, C., Merico, D., Hu, P., & Janitz, M. (2013). RNA-Seq analysis of the parietal cortex in Alzheimer's disease reveals alternatively spliced isoforms related to lipid metabolism. *Neuroscience Letters, 536*, 90–95. doi:10.1016/j.neulet.2012.12.042

Miyamoto, D. T., Zhu, H., Lee, R. J., Desai, N., Trautwein, J., Arora, K. S., Shioda, T., Broderick, K. T., Zheng, Y., Fox, D. B., Toner, M., Haber, D. A., Brannigan, B. W., Ramaswamy, S., Wu, C.-L., Kapur, R., Wittner, B. S., Sequist, L. V., Smith, M. R., … Desai, R. (2015). RNA-Seq of single prostate CTCs implicates noncanonical Wnt signaling in antiandrogen resistance. *Science, 349*(6254), 1351–1356. doi:10.1126/science.aab0917

Morán, I., Akerman, I., Van De Bunt, M., Xie, R., Benazra, M., Nammo, T., Arnes, L., Nakić, N., García-Hurtado, J., Rodríguez-Seguí, S., Pasquali, L., Sauty-Colace, C., Beucher, A., Scharfmann, R., Van Arensbergen, J., Johnson, P. R., Berry, A., Lee, C., Harkins, T., & Ferrer, J.. (2012). Human β cell transcriptome analysis uncovers lncRNAs that are tissue-specific, dynamically regulated, and abnormally expressed in type 2 diabetes. *Cell Metabolism, 16*(4), 435–448. doi:10.1016/j.cmet.2012.08.010

Moretto, A., & Colosio, C. (2011). Biochemical and toxicological evidence of neurological effects of pesticides: The example of Parkinson's disease. *NeuroToxicology, 32*(4), 383–391. doi:10.1016/j.neuro.2011.03.004

Morjaria, J. B., Caruso, M., Emma, R., Russo, C., & Polosa, R. (2018). Treatment of allergic rhinitis as a strategy for preventing asthma. *Current Allergy and Asthma Reports, 18*(4). doi:10.1007/s11882-018-0781-y

Moulton, P. V., & Yang, W. (2012). Air pollution, oxidative stress, and Alzheimer's disease. *Journal of Environmental and Public Health, 2012*, 1–9. doi:10.1155/2012/472751

Mouradian, M. M. (2012). MicroRNAs in Parkinson's disease. *Neurobiology of Disease, 46*(2), 279–284. doi:10.1016/j.nbd.2011.12.046

Mozaffarian, D., Benjamin, E. J., Go, A. S., Arnett, D. K., Blaha, M. J., Cushman, M., Das, S. R., de Ferranti, S., Després, J.-P., Fullerton, H. J., Howard, V. J., Huffman, M. D., Isasi, C. R., Jiménez, M. C., Judd, S. E., Kissela, B. M., Lichtman, J. H., Lisabeth, L. D., Liu, S., & Turner, M. B. (2015). Heart disease and stroke statistics 2016 update. In *Circulation*, 133 (4). doi:10.1161/cir.0000000000000350

Musgrove, R. E., Jewell, S. A., & Di Monte, D. A. (2015). Overview of neurodegenerative disorders and susceptibility factors in neurodegenerative processes. In Aschner, M., & Costa, L. G. (eds), *Environmental Factors in Neurodevelopmental and Neurodegenerative Disorders* (pp. 197–210). Amsterdam: Elsevier/Academic Press. doi:10.1016/B978-0-12-800228-5.00010-8

National Cancer Institute. (2018). *Cancer-Causing Substances in the Environment.* https://www.cancer.gov/about-cancer/causes-prevention/risk/substances

NINDS. (2014). Amyotrophic Lateral Sclerosis. In *Encyclopedia of the Neurological Sciences.* NIH Publication No. 16–916. doi:10.1016/B978-0-12-385157-4.00608-4

Nomura, S., Satoh, M., Fujita, T., Higo, T., Sumida, T., Ko, T., Yamaguchi, T., Tobita, T., Naito, A. T., Ito, M., Fujita, K., Harada, M., Toko, H., Kobayashi, Y., Ito, K., Takimoto, E., Akazawa, H., Morita, H., Aburatani, H., & Komuro, I. (2018). Cardiomyocyte gene programs encoding morphological and functional signatures in cardiac hypertrophy and failure. *Nature Communications, 9*(1), 1–17. doi:10.1038/s41467-018-06639-7

Nowotny, K., Jung, T., Höhn, A., Weber, D., & Grune, T. (2015). Advanced glycation end products and oxidative stress in type 2 diabetes mellitus. *Biomolecules, 5*(1), 194–222. doi:10.3390/biom5010194

Obeidat, M., Sadatsafavi, M., & Sin, D. D. (2019). Precision health: treating the individual patient with chronic obstructive pulmonary disease. *Medical Journal of Australia, i*, mja 2.50138. doi:10.5694/mja2.50138

Olden, K., & White, S. L. (2005). Health-related disparities: Influence of environmental factors. *Medical Clinics of North America, 89*(4), 721–738. doi:10.1016/j.mcna.2005.02.001

Paouri, E., & Georgopoulos, S. (2019). Systemic and CNS inflammation crosstalk: Implications for Alzheimer's Disease. *Current Alzheimer Research, 16.* doi:10.2174/1567205016666190321154618

Parachikova, A., Agadjanyan, M. G., Cribbs, D. H., Blurton-Jones, M., Perreau, V., Rogers, J., Beach, T. G., & Cotman, C. W. (2007). Inflammatory changes parallel the early stages of Alzheimer disease. *Neurobiology of Aging, 28*(12), 1821–1833. doi:10.1016/j.neurobiolaging.2006.08.014

Parrón, T., Requena, M., Hernández, A. F., & Alarcón, R. (2011). Association between environmental exposure to pesticides and neurodegenerative diseases. *Toxicology and Applied Pharmacology, 256*(3), 379–385. doi:10.1016/j.taap.2011.05.006

Pathak, U., Gupta, N. C., & Suri, J. C. (2019). Risk of COPD due to indoor air pollution from biomass cooking fuel: A systematic review and meta-analysis. *International Journal of Environmental Health Research, 00*(00), 1–14. doi:10.1080/09603123.2019.1575951

Pedrotty, D. M., Morley, M. P., & Cappola, T. P. (2012). Transcriptomic biomarkers of cardiovascular disease. *Progress in Cardiovascular Diseases, 55*(1), 64–69. doi:10.1016/j.pcad.2012.06.003

Poitout, V., Hagman, D., Stein, R., Artner, I., Robertson, R. P., & Harmon, J. S. (2006). Regulation of the insulin gene by glucose and fatty acids. *The Journal of Nutrition, 136*(4), 873–876. doi:10.1093/jn/136.4.873

Poy, M. N., Hausser, J., Trajkovski, M., Braun, M., Collins, S., Rorsman, P., Zavolan, M., & Stoffel, M. (2009). miR-375 maintains normal pancreatic alpha- and beta-cell mass. *Proc Natl Acad Sci U S A, 106*(14), 5813–5818. doi:10.1073/pnas.0810550106

Prendecki, M., & Dorszewska, J. (2014). The role of microRNA in the pathogenesis and diagnosis of neurodegenerative diseases. *Austin Alzheimer's and Parkinson's Disease, 1,* id1013.

Pringsheim, T., Jette, N., Frolkis, A., & Steeves, T. D. L. (2014). The prevalence of Parkinson's disease: A systematic review and meta-analysis. *Movement Disorders, 29*(13), 1583–1590. doi:10.1002/mds.25945

Qian, J., Wolters, F. J., Beiser, A., Haan, M., Ikram, M. A., Karlawish, J., Langbaum, J. B., Neuhaus, J. M., Reiman, E. M., Roberts, J. S., Seshadri, S., Tariot, P. N., Woods, B. M., Betensky, R. A., & Blacker, D.. (2017). APOE-related risk of mild cognitive impairment and dementia for prevention trials: An analysis of four cohorts. *PLOS Medicine, 14*(3), e1002254. doi:10.1371/journal.pmed.1002254

Qu, L., Ding, J., Chen, C., Wu, Z. J., Liu, B., Gao, Y., Chen, W., Liu, F., Sun, W., Li, X. F., Wang, X., Wang, Y., Xu, Z. Y., Gao, L., Yang, Q., Xu, B., Li, Y. M., Fang, Z. Y., Xu, Z. P., & Wang, L. H. (2016). Exosome-transmitted lncARSR promotes sunitinib resistance in renal cancer by acting as a competing endogenous RNA. *Cancer Cell, 29*(5), 653–668. doi:10.1016/j.ccell.2016.03.004

Ren, S., Wang, F., Shen, J., Sun, Y., Sun, Y., Xu, W., Lu, J., Wei, M., Xu, C., Wu, C., Zhang, Z., Gao, X., Liu, Z., Hou, J., & Huang, J. (2013). Long non-coding RNA metastasis associated in lung adenocarcinoma transcript 1 derived miniRNA as a novel plasma-based biomarker for diagnosing prostate cancer. *European Journal of Cancer, 49*(13), 2949–2959. doi:10.1016/j.ejca.2013.04.026

Ren, X., Kang, B., & Zhang, Z. (2018). Understanding tumor ecosystems by single-cell sequencing: Promises and limitations. *Genome Biology, 19*(1), 1–14. doi:10.1186/s13059-018-1593-z

Ricci, C., Marzocchi, C., & Battistini, S. (2018). MicroRNAs as biomarkers in amyotrophic lateral sclerosis. *Cells, 7*(11), 219. doi:10.3390/cells7110219

Ringman, J. M., & Coppola, G. (2013). New genes and new insights from old genes. *CONTINUUM: Lifelong Learning in Neurology, 19*(2), 358–371. doi:10.1212/01.CON.0000429179.21977.a1

Roggli, E., Britan, A., Gattesco, S., Lin-Marq, N., Abderrahmani, A., Meda, P., & Regazzi, R. (2010). Involvement of microRNAs in the cytotoxic effects exerted by proinflammatory cytokines on pancreatic β-cells. *Diabetes, 59*(4), 978–986. doi:10.2337/db09-0881

Saeedi Borujeni, M. J., Esfandiary, E., Baradaran, A., Valiani, A., Ghanadian, M., Codoñer-Franch, P., Basirat, R., Alonso-Iglesias, E., Mirzaei, H., & Yazdani, A. (2019). Molecular aspects of pancreatic β-cell dysfunction: Oxidative stress, microRNA, and long noncoding RNA. *Journal of Cellular Physiology, 234*(6), 8411–8425. doi:10.1002/jcp.27755

Saeedi Borujeni, M. J., Esfandiary, E., Taheripak, G., Codoñer-Franch, P., Alonso-Iglesias, E., & Mirzaei, H. (2018). Molecular aspects of diabetes mellitus: Resistin, microRNA, and exosome. *Journal of Cellular Biochemistry, 119*(2), 1257–1272. doi:10.1002/jcb.26271

Sauler, M., Lamontagne, M., Finnemore, E., Herazo-Maya, J. D., Tedrow, J., Zhang, X., Morneau, J. E., Sciurba, F., Timens, W., Paré, P. D., Lee, P. J., Kaminski, N., Bossé, Y., & Gomez, J. L. (2018). The DNA repair transcriptome in severe COPD. *European Respiratory Journal, 52*(4), 1701994. doi:10.1183/13993003.01994-2017

Sawyer, G., Miles, J., Lewis, S., Fitzharris, P., Pearce, N., & Beasley, R. (1998). Classification of asthma severity: Should the international guidelines be changed? *Clinical and Experimental Allergy, 28*(12), 1565–1570. doi:10.1046/j.1365-2222.1998.00451.x

Scott, L. J., Erdos, M. R., Huyghe, J. R., Welch, R. P., Beck, A. T., Wolford, B. N., Chines, P. S., Didion, J. P., Narisu, N., Stringham, H. M., Taylor, D. L., Jackson, A. U., Vadlamudi, S., Bonnycastle, L. L., Kinnunen, L., Saramies, J., Sundvall, J., Albanus, R. D. O., Kiseleva, A., & Parker, S. C. J. (2016). The genetic regulatory signature of type 2 diabetes in human skeletal muscle. *Nature Communications, 7.* doi:10.1038/ncomms11764

Segerstolpe, Å., Palasantza, A., Eliasson, P., Andersson, E. M., Andréasson, A. C., Sun, X., Picelli, S., Sabirsh, A., Clausen, M., Bjursell, M. K., Smith, D. M., Kasper, M., Ämmälä, C., & Sandberg, R. (2016). Single-cell transcriptome profiling of human pancreatic islets in health and type 2 diabetes. *Cell Metabolism, 24*(4), 593–607. doi:10.1016/j.cmet.2016.08.020

Seronde, M. F., Vausort, M., Gayat, E., Goretti, E., Ng, L. L., Squire, I. B., Vodovar, N., Sadoune, M., Samuel, J. L., Thum, T., Solal, A. C., Laribi, S., Plaisance, P., Wagner, D. R., Mebazaa, A., & Devaux, Y. (2015). Circulating microRNAs and outcome in patients with acute heart failure. *PLoS ONE, 10*(11), 1–14. doi:10.1371/journal.pone.0142237

Shaw, A., Bradley, M. D., Elyan, S., & Kurian, K. M. (2015). Tumour biomarkers: Diagnostic, prognostic, and predictive. *BMJ (Online), 351*(July), 2–5. doi:10.1136/bmj.h3449

Shcherbatykh, I., & Carpenter, D. O. (2007). The role of metals in the etiology of Alzheimer's disease. *Journal of Alzheimer's Disease, 11*(2), 191–205. doi:10.3233/JAD-2007-11207

Sherer, T. B., Betarbet, R., Testa, C. M., Seo, B. B., Richardson, J. R., Kim, J. H., Miller, G. W., Yagi, T., Matsuno-Yagi, A., & Greenamyre, J. T. (2003). Mechanism of toxicity in rotenone models of Parkinson's disease. *The Journal of Neuroscience: The Official Journal of the Society for Neuroscience, 23*(34), 10756–10764.

Shou, Q., Jin, L., Lang, J., Shan, Q., Ni, Z., Cheng, C., Li, Q., Fu, H., & Cao, G. (2019). Integration of metabolomics and transcriptomics reveals the therapeutic mechanism underlying paeoniflorin for the treatment of allergic asthma. *Frontiers in Pharmacology, 9*(January), 1–12. doi:10.3389/fphar.2018.01531

Sole, C., Arnaiz, E., Manterola, L., Otaegui, D., & Lawrie, C. H. (2019). The circulating transcriptome as a source of cancer liquid biopsy biomarkers. *Seminars in Cancer Biology, January,* 100–108. doi:10.1016/j.semcancer.2019.01.003

Steiling, K., Lenburg, M. E., & Spira, A. (2013). Personalized management of chronic obstructive pulmonary disease via transcriptomic profiling of the airway and lung. *Annals of the American Thoracic Society, 10*(SUPPL). doi:10.1513/AnnalsATS.201306-190AW

Syamlal, G., Doney, B., & Mazurek, J. M. (2019). *Chronic Obstructive Pulmonary Disease Prevalence Among Adults Who Have Never Smoked, by Industry and Occupation — United States, 2013 – 2017. 68*(13), 2013–2017.

Sygitowicz, G., Tomaniak, M., Blaszczyk, O., Koltowski, L., Filipiak, K. J., & Sitkiewicz, D. (2015). Circulating microribonucleic acids miR-1, miR-21 and miR-208a in patients with symptomatic heart failure: Preliminary results. *Archives of Cardiovascular Diseases, 108*(12), 634–642. doi:10.1016/j.acvd.2015.07.003

Tan, C., Liu, X., & Chen, J. (2018). Microarray analysis of the molecular mechanism involved in Parkinson's disease. *Parkinson's Disease, 2018*, 1–12. doi:10.1155/2018/1590465

Tang, F., Barbacioru, C., Wang, Y., Nordman, E., Lee, C., Xu, N., Wang, X., Bodeau, J., Tuch, B. B., Siddiqui, A., Lao, K., & Surani, M. A. (2009). mRNA-Seq whole-transcriptome analysis of a single cell. *Nature Methods, 6*(5), 377–382. doi:10.1038/nmeth.1315

Testa, C. M., Sherer, T. B., & Greenamyre, J. T. (2005). Rotenone induces oxidative stress and dopaminergic neuron damage in organotypic substantia nigra cultures. *Molecular Brain Research, 134*(1), 109–118. doi:10.1016/j.molbrainres.2004.11.007

Thomas, B. (2009). Parkinson's disease: From molecular pathways in disease to therapeutic approaches. *Antioxidants {&} Redox Signaling, 11*(9), 2077–2082. doi:10.1089/ars.2009.2697

Thomas, C. C., & Philipson, L. H. (2015). Update on diabetes classification. *Medical Clinics of North America, 99*(1), 1–16. doi:10.1016/j.mcna.2014.08.015

Toivonen, J. M., Manzano, R., Oliván, S., Zaragoza, P., García-Redondo, A., & Osta, R. (2014). MicroRNA-206: A potential circulating biomarker candidate for amyotrophic lateral sclerosis. *PLoS ONE, 9*(2), e89065. doi:10.1371/journal.pone.0089065

Toskala, E., & Kennedy, D. W. (2015). Asthma risk factors. *International Forum of Allergy and Rhinology, 5*(September), S11–S16. doi:10.1002/alr.21557

Trojsi, F., Monsurrò, M., & Tedeschi, G. (2013). Exposure to environmental toxicants and pathogenesis of amyotrophic lateral sclerosis: State of the art and research perspectives. *International Journal of Molecular Sciences, 14*(8), 15286–15311. doi:10.3390/ijms140815286

Twine, N. A., Janitz, K., Wilkins, M. R., & Janitz, M. (2011). Whole transcriptome sequencing reveals gene expression and splicing differences in brain regions affected by Alzheimer's disease. *PLoS ONE, 6*(1), e16266. doi:10.1371/journal.pone.0016266

United Nations. (2019). *Sustainable Development Goals*. https://sdgs.un.org/goals

Viswambharan, V., Thanseem, I., Vasu, M. M., Poovathinal, S. A., & Anitha, A. (2017). miRNAs as biomarkers of neurodegenerative disorders. *Biomarkers in Medicine, 11*(2), 151–167. doi:10.2217/bmm-2016-0242

Wadhwa, R., Aggarwal, T., Malyla, V., Kumar, N., Gupta, G., Chellappan, D. K., Dureja, H., Mehta, M., Satija, S., Gulati, M., Maurya, P. K., Collet, T., Hansbro, P. M., & Dua, K. (2019). Identification of biomarkers and genetic approaches toward chronic obstructive pulmonary disease. *Journal of Cellular Physiology, 9*(February), 1–21. doi:10.1002/jcp.28482

Wadhwani, A. R., Affaneh, A., Van Gulden, S., & Kessler, J. A. (2019). Neuronal apolipoprotein E4 increases cell death and phosphorylated tau release in alzheimer disease. *Annals of Neurology*, ana.25455. doi:10.1002/ana.25455

Waller, R., Goodall, E. F., Milo, M., Cooper-Knock, J., Da Costa, M., Hobson, E., Kazoka, M., Wollff, H., Heath, P. R., Shaw, P. J., & Kirby, J. (2017). Serum miR-NAs miR-206, 143-3p and 374b-5p as potential biomarkers for amyotrophic lateral sclerosis (ALS). *Neurobiology of Aging, 55*, 123–131. doi:10.1016/j.neurobiolaging.2017.03.027

Wang, R., Li, N., Zhang, Y., Ran, Y., & Pu, J. (2011). Circulating microRNAs are promising novel biomarkers of acute myocardial infarction. *Internal Medicine, 50*(17), 1789–1795. doi:10.2169/internalmedicine.50.5129

Wang, Y., Wang, O., Li, W., Ma, L., Ping, F., Chen, L., & Nie, M. (2015). Variants in vitamin D binding protein gene are associated with gestational diabetes mellitus. *Medicine (United States), 94*(40), 1–7. doi:10.1097/MD.0000000000001693

Wang, Z., Gerstein, M., & Snyder, M. (2009). RNA-Seq: A revolutionary tool for transcriptomics in western equatoria state. *Nature Reviews Genetics, 10*(1), 57.

Weisskopf, M. G., Morozova, N., O'Reilly, E. J., McCullough, M. L., Calle, E. E., Thun, M. J., & Ascherio, A. (2009). Prospective study of chemical exposures and amyotrophic lateral sclerosis. *Journal of Neurology, Neurosurgery {&} Psychiatry, 80*(5), 558–561. doi:10.1136/jnnp.2008.156976

Weng, W., Liu, N., Toiyama, Y., Kusunoki, M., Nagasaka, T., Fujiwara, T., Wei, Q., Qin, H., Lin, H., Ma, Y., & Goel, A. (2018). Novel evidence for a PIWI-interacting RNA (piRNA) as an oncogenic mediator of disease progression, and a potential prognostic biomarker in colorectal cancer. *Molecular Cancer, 17*(1), 1–12. doi:10.1186/s12943-018-0767-3

Wessel, J., Chu, A. Y., Willems, S. M., Wang, S., Yaghootkar, H., Brody, J. A., Dauriz, M., Hivert, M.-F., Raghavan, S., Lipovich, L., Hidalgo, B., Fox, K., Huffman, J. E., An, P., Lu, Y., Rasmussen-Torvik, L. J., Grarup, N., Ehm, M. G., Li, L., & Goodarzi, M. O. (2015). Low-frequency and rare exome chip variants associate with fasting glucose and type 2 diabetes susceptibility. *Nature Communications, 6*(1). doi:10.1038/ncomms6897

WHO. (2017). *Preventing Noncommunicable Diseases (NCDs) by Reducing Environmental Risk Factors*. Geneva: World Health Organization. https://apps.who.int/iris/handle/10665/258796

WHO. (2018). *Diabetes*. Geneva: World Health Organization. https://www.who.int/health-topics/diabetes

WHO. (2019a). *Cancer*. Geneva: World Health Organization. https://www.who.int/health-topics/cancer#tab=tab_1

WHO. (2019b). *Noncommunicable Diseases*. Geneva: World Health Organization. https://www.who.int/health-topics/noncommunicable-diseases

Widera, C., Gupta, S. K., Lorenzen, J. M., Bang, C., Bauersachs, J., Bethmann, K., Kempf, T., Wollert, K. C., & Thum, T. (2011). Diagnostic and prognostic impact of six circulating microRNAs in acute coronary syndrome. *Journal of Molecular and Cellular Cardiology, 51*(5), 872–875. doi:10.1016/j.yjmcc.2011.07.011

Willett, W. C. (2002). Balancing life-style and genomics research for disease prevention. *Science, 296*(5568), 695–698. doi:10.1126/science.1071055

Williams, A. H., Valdez, G., Moresi, V., Qi, X., McAnally, J., Elliott, J. L., Bassel-Duby, R., Sanes, J. R., & Olson, E. N. (2009). MicroRNA-206 delays ALS progression and promotes regeneration of neuromuscular synapses in mice. *Science, 326*(5959), 1549–1554. doi:10.1126/science.1181046

World Health Organization. (2017a). *Cardiovascular Diseases (CVDs)*. Geneva: World Health Organization. https://www.who.int/health-topics/cardiovascular-diseases

World Health Organization. (2017b). *Chronic Obstructive Pulmonary Disease (COPD)*. Geneva: World Health Organization. https://www.who.int/news-room/fact-sheets/detail/chronic-obstructive-pulmonary-disease-(copd)

World Health Organization. (2018). *The Top 10 Causes of Death*. Geneva: World Health Organization. https://www.who.int/news-room/fact-sheets/detail/the-top-10-causes-of-death

World Health Organization. (2019). *Chronic Respiratory Diseases Retrieved*. Geneva: World Health Organization. https://www.who.int/health-topics/chronic-respiratory-diseases#tab=tab_1

Wu, F., Guo, Y., Chatterji, S., Zheng, Y., Naidoo, N., Jiang, Y., Biritwum, R., Yawson, A., Minicuci, N., Salinas-Rodriguez, A., Manrique-Espinoza, B., Maximova, T., Peltzer, K., Phaswanamafuya, N., Snodgrass, J. J., Thiele, E., Ng, N., & Kowal, P. (2015). Common risk factors for chronic non-communicable diseases among older adults in China, Ghana, Mexico, India, Russia and South Africa: The study on global AGEing and adult health (SAGE) wave 1. *BMC Public Health, 15*(1), 1–13. doi:10.1186/s12889-015-1407-0

Xia, W., Qiu, M., Chen, R., Wang, S., Leng, X., Wang, J., Xu, Y., Hu, J., Dong, G., Xu, P. L., & Yin, R. (2016). Circular RNA has-circ-0067934 is upregulated in esophageal squamous cell carcinoma and promoted proliferation. *Scientific Reports, 6*(September), 1–9. doi:10.1038/srep35576

Xin, Y., Kim, J., Okamoto, H., Ni, M., Wei, Y., Adler, C., Murphy, A. J., Yancopoulos, G. D., Lin, C., & Gromada, J. (2016). RNA sequencing of single human islet cells reveals type 2 diabetes genes. *Cell Metabolism, 24*(4), 608–615. doi:10.1016/j.cmet.2016.08.018

Yan, B., Tao, Z. F., Li, X. M., Zhang, H., Yao, J., & Jiang, Q. (2014). Aberrant expression of long noncoding RNAs in early diabetic retinopathy. *Investigative Ophthalmology and Visual Science, 55*(2), 941–951. doi:10.1167/iovs.13-13221

Yang, F., Liu, D. Y., Guo, J. T., Ge, N., Zhu, P., Liu, X., Wang, S., Wang, G. X., & Sun, S. Y. (2017). Circular RNA circ-LDLRAD3 as a biomarker in diagnosis of pancreatic cancer. *World Journal of Gastroenterology, 23*(47), 8345–8354. doi:10.3748/wjg.v23.i47.8345

Yates, J. R., Ruse, C. I., & Nakorchevsky, A. (2009). Proteomics by mass spectrometry: approaches, advances, and applications. *Annual Review of Biomedical Engineering, 11*, 49–79. doi:10.1146/annurev-bioeng-061008-124934

Yegambaram, M., Manivannan, B., Beach, T., & Halden, R. (2015). Role of environmental contaminants in the etiology of Alzheimer's disease: A review. *Current Alzheimer Research, 12*(2), 116–146. doi:10.2174/1567205012666150204121719

Yick, C. Y., Zwinderman, A. H., Kunst, P. W., Gruñberg, K., Mauad, T., Dijkhuis, A., Bel, E. H., Baas, F., Lutter, R., & Sterk, P. J. (2013a). Transcriptome sequencing (RNA-Seq) of human endobronchial biopsies: Asthma versus controls. *European Respiratory Journal, 42*(3), 662–670. doi:10.1183/09031936.00115412

Yick, C. Y., Zwinderman, A. H., Kunst, P. W., Gruñberg, K., Mauad, T., Fluiter, K., Bel, E. H., Lutter, R., Baas, F., & Sterk, P. J. (2013b). Glucocorticoid-induced changes in gene expression of airway smooth muscle in patients with asthma. *American Journal of Respiratory and Critical Care Medicine, 187*(10), 1076–1084. doi:10.1164/rccm.201210-1886OC

Zelensky, A. N., & Gready, J. E. (2005). The C-type lectin-like domain superfamily. *FEBS Journal, 272*(24), 6179–6217. doi:10.1111/j.1742-4658.2005.05031.x

Zhang, R., Xia, Y., Wang, Z., Zheng, J., Chen, Y., Li, X., Wang, Y., & Ming, H. (2017a). Serum long non coding RNA MALAT-1 protected by exosomes is up-regulated and promotes cell proliferation and migration in non-small cell lung cancer. *Biochemical and Biophysical Research Communications, 490*(2), 406–414. doi:10.1016/j.bbrc.2017.06.055

Zhang, X., Yang, R., Hu, B.-L., Lu, P., Zhou, L.-L., He, Z.-Y., Wu, H.-M., & Zhu, J.-H. (2017b). Reduced circulating levels of miR-433 and miR-133b are potential biomarkers for Parkinson's disease. *Frontiers in Cellular Neuroscience, 11.* doi:10.3389/fncel.2017.00170

Zhang, Y. (2018). Long non-coding RNA MALAT1 expression in patients with gestational diabetes mellitus. *International Journal of Gynecology & Obstetrics, 140*(2), 164–169.

Zhao, Y., Hu, J., Tan, Z., Liu, T., Zeng, W., Li, X., Huang, C., Wang, S., Huang, Z., & Ma, W. (2019). Ambient carbon monoxide and increased risk of daily hospital outpatient visits for respiratory diseases in Dongguan, China. *Science of the Total Environment, 668,* 254–260. doi:10.1016/j.scitotenv.2019.02.333

Zhou, S. S., Jin, J. P., Wang, J. Q., Zhang, Z. G., Freedman, J. H., Zheng, Y., & Cai, L. (2018). MiRNAS in cardiovascular diseases: Potential biomarkers, therapeutic targets and challenges review-article. *Acta Pharmacologica Sinica, 39*(7), 1073–1084. doi:10.1038/aps.2018.30

Zhou, Y., Chung, A. C. K., Fan, R., Lee, H. M., Xu, G., Tomlinson, B., Chan, J. C. N., & Kong, A. P. S. (2017). Sirt3 deficiency increased the vulnerability of pancreatic beta cells to oxidative stress-induced dysfunction. *Antioxidants & Redox Signaling, 27*(13), 962–976. doi:10.1089/ars.2016.6859

5

The Impact of MicroRNAs on Human Diseases

Rodolfo Iván Valdez Vega
Universidad de Guadalajara, Guadalajara, Mexico

Jorge Montiel Montoya
Instituto Politécnico Nacional–CIIDIR Sinaloa, Guasave, Mexico

José Luis Acosta Rodríguez
Instituto Politécnico Nacional–CIIDIR Sinaloa, Guasave, Mexico

CONTENTS

5.1 Introduction

The technologies that help sequence the human genome mainly focus on the protein-coding gene parts, also called exons. Because of this, sequencing results are often limited. The sequences that are responsible for protein coding are the most studied and best understood part of the human genome. However, the sequences referred to as "non-coding sequences" constitute an important fraction of the genome. The transcription of the human genome participates only in a small proportion (~3%) for protein coding. Hence, an issue is raised on the role of the non-protein-coding part of our genome, along with ribosomal RNAs (rRNAs) and transfer RNAs (tRNAs), involvement in other processes. Therefore, it is important to know the functions of some non-protein-coding transcripts (Perenthaler et al., 2019; Hess et al., 2020).

Due to certain questions about non-coding sequences, next-generation sequencing (NGS) technologies have been used to reveal mainly the composition of non-coding RNAs present in cell nuclei to understand the mechanisms by which non-coding RNAs are involved in biological processes and gene expression (O'Neill 2020). Sequencing technologies have been very useful for the discovery of small RNAs that do not directly participate in protein coding. These non-coding RNAs are known as microRNAs (miRNAs). Techniques such as RNA-seq have made it possible to quantify both over-expressed and under-abundant miRNAs, and this also allows detection of miRNA variants (Kappel and Keller 2017).

This chapter shows an account of the works done on human miRNAs, highlighting the use of various sequencing technologies such as NGS involving RNA-seq. In addition, other techniques that also have been used to study these molecules include microarrays and RT-qPCR, which are widely used to study miRNAs. These technologies have a number of differences in terms of reliability of results, accuracy, and costs of use. The RT-qPCR is one of the most used methods for the identification of miRNAs, as it allows reliable detection and measurement of the products generated during each cycle of the PCR process. The use of RNA-seq has also been described as an important technology, since it allows sequencing of large fragments of nucleic acids, genome assembly, and annotation methods. This technique has been used to classify a wide variety of nuclear RNAs, including miRNAs. Other emerging technologies, such as the Nanostring technique, are promising tools for RNA profiling. With this technique, miRNAs have been explored by detecting specific nucleic acid molecules from low amounts of starting material without the need for reverse transcription or cDNA amplification (Kappel and Keller 2017).

5.2 miRNAs

Regulation of gene expression by small RNAs has been found in all three domains of life (bacteria, archaea, and eukaryotes), having evolved primarily as a defense against foreign nucleic acids. These RNA-based interference systems consist mainly of two components, a nucleic acid that recognizes the specific sequence of the target, and an effector protein that helps mediate downstream effects (Dexheimer and Cochella 2020).

These small RNAs have been described as microRNAs (miRNAs). These are single-stranded RNAs and are referred to as micro because their size ranges from 19 to 25 nucleotides in length, and these are generated in endogenous hairpin transcripts. The miRNAs function as a guide in post-transcriptional gene silencing because they can recognize and bind to messenger RNA (mRNA) transcripts at a certain specific site in the sequence. In general, miRNAs participate in the inhibition of protein synthesis by either repressing translation and/or causing the removal of the poly(A) tail (deadenylation) and subsequent degradation of mRNA targets. However, certain types of miRNAs have also been reported to participate in mRNA translation activation processes (Bartel 2018; Alemán-Ávila et al., 2019; Mori et al., 2019).

The miRNA database (miRBase) lists 38,589 records representing miRNA hairpin precursors from 271 species. These hairpin precursors help create approximately 48,860 mature miRNA sequences. About 1917 hairpin precursors and 2654 mature sequences have been found in the human genome. The miRNAs also influence the control of different biological processes involved in homeostasis, leading to important consequences in normal development and physiology, disease, and evolution (Kozomara et al., 2019; Mori et al., 2019).

Another characteristic of miRNAs is their vital role in development, cell differentiation, apoptosis, innate immunity, and molecular metabolism is based mainly on post-transcriptional negative regulation, due to their high dynamics and efficiency, as they can regulate more than 100 target genes, while more than 50% of mRNA transcripts can be upregulated by more than two miRNAs (Plotnikova et al., 2019).

A particular microRNA can target a variety of different mRNAs, and a particular mRNA can bind to a variety of microRNAs, either simultaneously or in a dependent manner. These microRNA-mRNA interactions are still far from being fully understood, but as miRNAs are further studied, their complexity in various functions and the multiple factors that tightly control their processing and activities have been demonstrated. Not surprisingly, the biogenesis and function of miRNAs are tightly regulated, and their dysregulation is often associated with human diseases, such as cancer and neurodevelopmental disorders (Cazzanelli and Wuertz-Kozak 2020; Gao et al., 2020).

MicroRNAs can be classified depending on their location in two ways: (i) those that have action inside the cell are called intracellular miRNAs, and (ii) circulating miRNAs (c-miRNAs) that are found in different body fluids and function as regulators of gene expression and mediators of intercellular communication (Sanjurjo et al., 2016).

5.3 Biogenesis of miRNAs

5.3.1 Overview of miRNA Biogenesis and Function

The biogenesis of miRNAs begins with the processing of RNA polymerase II/III transcripts in a post- or co-transcriptional manner. Of the miRNAs currently identified, about half are intragenic (i.e., processed from introns) and few are processed from exons of protein-coding genes, while the remainder are intergenic. Primary miRNA transcripts, or Pri-miRNAs, contain cap structures as well as poly(A) tails. These are unique properties of RNA polymerase II class gene transcripts (O'Brien et al., 2018; Pong and Gullerova 2018).

The miRNAs are processed from a common primary transcript; however, these long pri-miRNAs are separated by a complex called a "microprocessor", the separation of which is a key step in regulating miRNA biogenesis. The differential activity of the complex is an important determinant of steady-state miRNA levels and demonstrates the independent control of the activities involved in miRNA biogenesis (Jiang and Yan 2016; Balasubramanian et al., 2020). Two proteins called DROSHA and DGCR8 together form the crucial complex for the processing (microprocessor) of pri-miRNAs in the nucleus, which is the first step of miRNA maturation. The DROSHA is a type III RNase, which has two RNase III domains in sequence that recognize the structural pattern that is produced by intramolecular base pairing of miRNAs, termed stem-loop structures. (Lin and Gregory 2015). The DGCR8 associates with the apical loops of dsRNA (double-stranded RNA) and the heme RNA-binding domain, respectively, which allows for more efficient and precise processing. Hemin mainly stimulates the expression of pri-miRNAs possessing the UGU motif, allowing miRNAs to differentiate at maturation (Nguyen et al., 2018). The DROSHA and DGCR8, respectively, interact with the basal UG motif and basal and apical UGU; this aids their targeting on the pri-miRNA at a position where DROSHA and DGCR8 localize to the basal and apical junctions of the pri-miRNA, respectively. CNNC interacts with SRSF3 (SRp20) to stimulate such a complex for the elaboration of pri-miR-NAs (Denzler and Stoffel 2015; Nguyen et al., 2015; Kim et al., 2018).

The "microprocessor" (Drosha-DGCR8) also requires association to a histone H1-like chromatin protein (HP1BP3). This protein binds specifically to endogenous pri-miRNAs, which aids Drosha/pri-miRNA association. Deletion of HP1BP3 risks the production and processing of pri-miRNAs by causing early release of pri-miRNAs from chromatin. It is suggested that HP1BP3 promotes co-transcriptional processing of miRNAs by retaining the created pri-miRNA transcripts in chromatin (Liu et al., 2016). Once the complex forms the pre-miRNAs, they are exported directly to the cytoplasm by a complex called exportin 5 (XPO5)/RanGTP in the case of the canonical pathway (O'Brien et al., 2018). The already exported pre-miRNAs are processed by an RNaseIII

Dicer, which plays a key role in the generation of miRNAs via the RNA inter-ference (RNAi) pathway. This processes a large diversity of double-stranded RNA precursors (dsRNA) to generate microRNA (miRNA) products of ~22 nucleotides (nt). Several Dicer-associated proteins, such as TRBP, PACT, and ADAR1, are known (Liu et al., 2018b; Matsuyama and Suzuki 2019).

Humans express only one type of Dicer, and its main function is to pro-cess pre-miRNAs into shorter-length intermediate miRNAs, which are still double-stranded. This complex is formed by an amino-terminal RNA helicase, which is interrupted by a helical insertion called Hel-I, a platform domain and a PAZ domain. Catalytic activity is provided by RIIIDa and RIIIDb (Treiber et al., 2019).

Once the miRNA duplex is obtained, an Argonaute (AGO) protein comes into operation. It has been described to influence the silencing or transcrip-tional activation of miRNA genes (~22 nt) and can be associated with other AGO proteins to induce translational inhibition or exonucleolytic degradation of mRNA of specific transcripts, such as small interfering RNAs (siRNAs) and miRNAs. In humans, four types of AGO (AGO1–4) are expressed. The AGO2 is the most relevant for its participation, and it has even been inferred that it is the only member of the Argonaute proteins that functions as an mRNA cutting tool, due to its unique structure (Müller et al., 2020). MicroRNAs and RNAs make an assembly with AGO proteins in the RNA-induced silencing complex called RISC complex. This complex mediates sequence-specific tar-get gene silencing. The binding that exists between RISC, miRNA and AGO, is not just a simple binding but follows an orderly multi-step pathway that requires specific accessory factors. Some steps of RISC assembly and RISC-mediated gene silencing depend on or are facilitated by specific intracellular platforms, suggesting their spatial regulation. This anchoring with RISC in which miRNA duplexes are loaded onto the protein AGO results in a com-plex containing AGO and a small RNA duplex called "pre-RISC". Maturation of this complex can be divided into two simple steps: coinage and passen-ger ejection. In wedging, one end of the duplex is opened by the N domain of AGO. The last step is passenger ejection, in which the two small RNA strands proceed to separate, and the passenger strand detaches from AGO. The complex formed by these AGO-bound strands and the single-stranded guidewire is called "mature RISC" (Kobayashi and Tomari 2016; Daugaard and Hansen 2017; Wu et al., 2020).

5.4 Mechanisms of miRNA-Mediated Gene Regulation

The so-called mature miRNAs are more stable RNA molecules compared to mRNAs. The complex formed by the AGO protein and the miRNA forms the microRNA-induced silencing complex (miRISC), which is essential to execute the miRNA function. The miRISC pursuit of a specific target is due to interaction with reverse complementary sequences within the 3′ untranslated region (UTR) of the target RNAs, which are known as miRNA response elements (MREs). The complementarity of MREs gives rise to two processes: an AGO2-dependent cleavage of the target mRNA or a translational inhibition carried out by miRISC and a decay of the target mRNA. When miRNA-MRE interact in a complementary manner, they induce AGO2 endonuclease activity and cleavage of the target mRNA. However, such interaction weakens the AGO-end 3′ binding of miRNA which promotes its degradation (O'Brien et al., 2018; Stavast and Erkeland 2019). The glycine-tryptophan family proteins (GW182) also participate in the silencing process. They interact directly with AGO proteins. The TNRC6A protein is part of this family of proteins, and it has functions in translational repression, poly(A) shortening in mRNAs, and decapping (Park et al., 2017). In addition, AGO2 helps to silence these targets through cleavage activity. To perform cleavage, the miRNA and the target mRNA are required to perfectly match in the miRNA core region. AGO3 also helps to cleave target RNAs under certain circumstances (Nishi et al., 2015).

5.5 Expression of Disease-Associated miRNAs

5.5.1 Cancer

Through the gene regulation mentioned in the previous section, miRNAs play an important role in various cellular processes, including proliferation, cell cycle control, programmed cell death, differentiation, invasion, and tissue-specific functions such as immune responses, hormone secretion, and angiogenesis. These processes are known to be associated with the development and progression of cancer (Iqbal et al., 2019). Genome-wide studies have shown that irregular expression of miRNAs is associated with various types of cancer through different mechanisms, malfunction of the miRNA biogenesis machinery, amplification/deletion of the miRNA encompassing region, or transcriptional deregulation (Zhang et al., 2019). The oncogenic or tumor suppressor role of miRNAs in cancer depends mainly on their inhibitory targets. These can be classified as follows: tumor suppressors (i.e., they help to suppress oncogenes), and those that target tumor suppressors that are potentially oncogenic, are known as oncomir. Several studies have reported the low expression of Drosha and Dicer in some types of cancer, and in contrast, other studies also suggest the overregulation of Drosha, Dicer, DGCR8, XPO5, and AGO2 in other types of cancer (Chen et al., 2020b).

One of the first oncomirs found to be overexpressed in various types of cancer, such as gliomas, breast cancer, and colorectal cancer, was miR-21. Some of its cancer-associated target genes are phosphatase and tensin homolog (PTEN), programmed cell death protein 4 (PDCD4), cysteine-rich reversion-inducing protein with Kazal motifs (RECK), and activator of transcription signaling 3 (STAT3) (Bautista-Sánchez et al., 2020). The miR-346 is also associated with upregulation by its overexpression in the ascending colon and may be responsible for suppression of cellular protective responses modulated by VDR and TNF-α. This situation may lead to inadequate suppression of neoplasia. The miR-1246 is also considered an oncomir in several types of cancer, mainly found enriched in exosomes derived from human cancer cells. This exosomal miRNA is generated independently of Drosha and Dicer, suggesting that its biogenesis is through the non-canonical pathway. There are other oncomirs that have been studied for their dual function such as miR-186, which is a tumor suppressor miRNA, while conflicting reports have verified that miR-186 is an oncomir (Xu et al., 2019; Kempinska-Podhorodecka et al., 2020; Xiang et al., 2020).

The miR-18a is another case of miRNA having dual function, which can act as both oncogene and suppressor. It is a member of the primary transcript called miR-17-92a. The miR-18a regulates a wide range of genes that are involved in proliferation, cell cycle, apoptosis, response to different types of stress, autophagy, and differentiation (Kolenda et al., 2020).

The miRNA genes, when transcribed by RNA polymerase II, are subject to the same regulatory methods as protein-coding genes. Epigenetic dysregulation of miRNAs related to CpG methylation of the promoters of two gene-encoding members of the miR-200 family has been reported. This family of miRNAs is associated with breast and colorectal cancer progression. Some regulators of miRNAs in cancer inhibit their expression; this is due to low expression levels of Drosha and Dicer, leading to low miRNA expression in cancer cells. The DDX17 is a cofactor of the miRNA "microprocessor" complex. DDX17 irregularity is also associated with cancer development, and another factor involved is AGO2 phosphorylation, which in turn decreases Dicer binding to AGO2, inhibiting the process of miRNA biogenesis from precursor to mature miRNA (Wu 2020; Goodall and Wickramasinghe 2021).

In ovarian cancer, various expression profiles of microRNA have been evaluated. The miRNAs function through degradation or inhibition of translation of the target mRNA. Several miRNAs, including miR-135a-3p, miR-200c, miR-216a and miR-340, are associated with the regulation of epithelial-mesenchymal transit, which helps modulate the invasiveness of cancer cells (Ghafouri-Fard et al., 2020). The miR-205 was also found to be overexpressed in ovarian cancer, and its expression is associated with the development of metastasis. This ovarian cancer-derived miRNA is associated with exosomes and the functioning of lipid rafts, where it plays an important role in the regulation of the exosomal miR-205 capacity (He et al., 2019).

5.5.2 Neurodegenerative Diseases

The miRNAs have been related as potential biomarkers for central nervous system (CNS) disorders as they are mainly related to the adjustment of developmental processes and cell differentiation. They are also extremely stable, are mainly found in tissues, and can be quantitatively determined. The miR-29 family, miR-34a-5p, and miR-132-3p are regarded as common deregulated circulating miRNAs found in CNS disorders (Konovalova et al., 2019).

Most neurodegenerative disorders are mainly associated with oxidative stress. Such stress is related to expression and regulation of multiple microRNAs, thus influencing processes such as neurodegeneration, mitochondrial dysfunction, proteostasis dysregulation and increased neuroinflammation (van den Berg et al., 2020).

Harmful exposure to various environmental factors throughout life generates damage to cells or they are affected to the point of aging. Factors such as DNA methylation, histone modification, and non-coding RNAs influence these processes. The miRNAs have been studied in this regard. Families such as miR-34, miR-29, and miR-126 are involved in both neurotoxicant-induced neurodegeneration and aging. The miR-34 family members demonstrate the possibility of the connection between miRNAs aging and neurotoxicant-induced neurodegeneration (Singh and Yadav 2020).

Several microRNAs have been found to directly decrease the biosynthesis of certain proteins present in neuronal tissue membranes (e.g., Amyloid precursor protein, APP). The APP is mainly involved in synapse formation and neuronal plasticity. The miR-101 can target APP mRNA and consequently suppress APP production. The miR-16 was also found to target APP mRNA and suppress APP levels in Alzheimer's disease (Idda et al., 2018).

5.5.3 Cardiovascular Diseases

In the cardiovascular system, miRNAs help to regulate the functions of various cells, such as cardiomyocytes, endothelial cells, smooth muscle cells and fibroblasts. The study of miRNAs in the cardiovascular system generates new insight into various disorders such as myocardial infarction, hypertrophy, fibrosis, heart failure, arrhythmia, inflammation, and atherosclerosis. The miRNAs may be overexpressed in diseased tissues, and these may be released into the circulation, in addition to being manipulated by various factors that may affect the course of a disease (Wojciechowska et al., 2017).

Several miRNAs have been implicated in the regulation of myocardial remodeling resulting from ischemia/reperfusion (I/R) injury in ischemic heart disease. The expression of miR-762 and miR-210 (see Table 5.1) were overexpressed when a myocardial infarction event occurred, conversely to

TABLE 5.1

miRNAs Associated with Cardiovascular Diseases

miRNA	Regulation	Type
miR-762 and miR-210	Up-regulated	Myocardial infarct
miR-1	Down-regulated	Myocardial infarct
miR-21	Up-regulated	Myocardial infarct
miR-122	Up-regulated	Hypertension
miR-10a, miR-672-5p, miR-139-5p, miR-135b, miR-142-3p, miR-150, and miR-99a	—	Cardiac hypertrophy

miR-1, which was inhibited in the same event. Recently, miR-762 was found to be highly translocated to mitochondria and markedly upregulated by hypoxia/reoxygenation in cardiomyocytes (Song et al., 2019). Expression of miR-21 and miR-223 has also been implicated in infarcts. These are mainly expressed in the border zone of infarcted hearts but significantly reduced in the non-infarcted zone. Overexpressed miR-21 protects against I/R injury by reducing myocardial infarct size and apoptosis, via its target genes, PTEN and PDCD4. It has also been observed that miR-21 expression was inhibited and autophagy was markedly increased in H9c2 cells during heart attack injury (Huang et al., 2017; Liu et al., 2018a; Samidurai et al., 2018).

Another of the abnormalities related to cardiovascular diseases is hypertension. Hypertension is a complex cardiovascular syndrome, and its pathogenesis may be directly related to abnormalities in miRNAs. Specifically, miR-122 is recognized to participate in the regulation of cardiovascular fibrosis and endothelial function during hypertension (Liu et al., 2020).

Various miRNAs, including miR-10a, miR-672-5p, miR-139-5p, miR-135b, miR-142-3p, miR-150, and miR-99a, are related to the regulation of cardiac hypertrophy. Particularly, miR-99a is involved in negatively regulating physiological hypertrophy through mTOR signaling. This may provide a new therapeutic approach for pressure overload heart failure (Li et al., 2016; Wehbe et al., 2019).

5.5.4 Metabolic Syndromes

Metabolic syndromes can be clinically classified as a complex and varied set of diseases including obesity and diabetes mellitus. Recent studies have found that cells secrete nanoscale vesicles containing proteins, lipids, nucleic acids, and membrane receptors, which are involved in signal transduction processes and the transport of materials to neighboring and distant cells. The miRNAs associate with vesicles called exosomes to regulate physiological function and pathological processes of metabolic disorders (Yao et al., 2018).

Several studies have found certain miRNA families in urinary and serum fluid in patients with type 1 and 2 diabetes. Most of these studies simply report cross-sectional associations of urinary miRNAs with albuminuria status (Colhoun and Marcovecchio 2018).

The miR-375 is considered an essential miRNA for normal glucose homeostasis, β-cell proliferation and α-cell turnover. In addition, it has been identified as a pancreatic islet cell-specific miRNA that targets myotrophin mRNA. Various miRNAs such as miR-200, miR-126, miR-21, miR-29, miR-7, miR-3666, and miR-135a have also been specifically identified in type 2 diabetes, whereas miR-326 and miR-146 have been described in type 1 diabetes (Aghaei Zarch et al., 2020).

Early stages of diabetes often result in endothelial dysfunction and microvascular rarefaction. Insufficient myocardial angiogenesis is the main manifestation of ischemic cardiovascular disease due to diabetes; significantly increased fluxes of miR-320-3p-loaded exosomes have been detected in cardiac cell culture in diabetic patients (Beuzelin and Kaeffer 2018).

Diseases associated with metabolic syndromes include the diabetic retinopathy (DR), which is linked to diabetes. The molecules miR-221, SIRT1, and Nrf2 are associated with apoptosis and proliferation, and their expression is altered in patients with DR (Chen et al., 2020a).

Obesity is part of the so-called metabolic syndromes, and it is considered an important risk factor for several metabolic diseases. The metabolism of individuals with obesity has been observed to generate changes in the expression profile of miRNAs. It has been confirmed that most miRNAs are inhibited whereas most proteins are overexpressed in obese subjects (Huang et al., 2018). One miRNA associated with obesity is miR-374a-5p, which has been shown to be overexpressed in morbidly obese subjects and appears to be linked with inhibitory regulation of proinflammatory markers that are related to insulin resistance (Doumatey et al., 2018).

The miRNAs family called miR-33 are mainly associated with obesity and have been described as an important regulator of lipid metabolism by targeting a series of genes involved in reverse cholesterol transport (RCT). Moreover, in the control of lipid metabolism, miR-33 has been described to target genes involved in other important metabolic functions, such as fatty acid metabolism (CPT1, CROT, HADHβ), insulin signaling (IRS2), and mitochondrial function (AMPK, PGC1α). The ability of miR-33 to regulate different functions suggests that miR-33 may have metabolic activity in different tissues. This is especially interesting because miR-33a and miR-33b are intronic miRNAs encoded within the sterol regulatory-element binding proteins (SREBP) genes SREBP-2 and SREBP-1 (SREBP). These factors play an important role in the cellular metabolism of cholesterol and fatty acids, and its presence has been described in several different metabolic tissues under conditions of obesity and insulin resistance (Ouimet et al., 2017; Price et al., 2018).

5.6 Outlook

5.6.1 Prediction of Targets and the Regulatory Mechanism of Extracellular miRNAs

The miRNAs play a fundamental role in intercellular and inter-tissue communication; this is strongly supported by the fact that miRNAs use a vesiculation mechanism to be exported and imported by cells. The presence of miRNAs has also been demonstrated in different body fluids, and their levels were associated with disease progression. Since then, the mechanisms of extracellular transport have become the subject of research. It is known that this mechanism occurs through two main routes: (a) active transport through extracellular vesicles, and (b) transport as part of protein-miRNA complexes (Mori et al., 2019).

To gain insight into the origin and regulatory mechanism of these miRNAs, bioinformatics tools have been developed to focus specifically on miRNA biogenesis processes to help find evidence for the biology of miRNAs. Annotation tools are among the most important devices in this field. A data storage platform is required, miRBase which is the main portal for miRNA storage and acts as a repository. This database collects all known miRNA sequences and annotations of known miRNAs for all species (Chen et al., 2019a).

To identify the function of miRNAs and their involvement in biological processes, it is necessary to identify the target genes of miRNAs, which will greatly help to understand their biological functions. Computational approaches, which excel in genomic data management transcriptomics and proteomics, should provide tools of the relative position of miRNAs in various biological networks. This is often best achieved by computational prediction followed by experimental validation of these miRNA-miRNA interactions, as it is a complex handling of information. Currently, several computational analysis technologies exist that use different approaches for predicting both miRNAs and their targets. As a result, machine learning (ML) models based on predictions and models based on deep neural networks have been successfully developed. The proper integration of these prediction algorithms can significantly improve the prediction accuracy and information on miRNA interaction with the target gene. This issue requires much more complex computational approaches that not only predict the target gene accurately, but also predict the regulatory networks of miRNAs and model the interaction between miRNAs (Parveen et al., 2019; Riolo et al., 2020; Schäfer and Ciaudo 2020).

These tools are designed to identify new miRNA-disease associations. The identification of miRNA disease has become a target of interest in biomedical research. Understanding this association will accelerate the identification of disease pathogenesis at the molecular level and will help the design of molecular tools for the diagnosis, treatment, and prevention of such diseases. For this, computational tools are used that could select the most promising miRNA-disease pairs for experimental validation and significantly reduce the time and cost of biological experiments (Chen et al., 2019b).

5.6.2 miRNAs as Biomarkers

Biomarker is a term that defines different types of objective indicators whether present in health or disease. These indicators have become increasingly accurate and reliable. Today they are referred to as certain molecules, usually proteins, which are detected in different body fluids, through specific means in medical laboratories. A potential advantage of miRNAs lies in their potential to be used as multimarker models for more accurate diagnosis, guided treatment, and response to treatment. For example, the identification of different urinary miRNA families of lupus nephritis has promoted early detection of renal fibrosis (Condrat et al., 2020). In cancer, information already exists on miRNAs and their roles as potential oncogenes or tumor suppressors (Table 5.2). Overexpression of these molecules was correlated with the occurrence of many cancer diseases, and therefore they are considered a molecular tool for non-invasive cancer diagnosis and prognosis. About 31 different miRNAs with potential biomarker for both diagnosis and response to treatment have been reported, as their deregulated expression before therapy returned to normal after receiving treatment (Filipów and Łaczmański 2019).

TABLE 5.2

miRNAs Associated with Cancer

miRNA	Regulation	Type	Disease
miR-21	Up-regulated	Oncomir	Gliomas, breast cancer, colorectal cancer
miR-346	Up-regulated	Oncomir	Colorectal cancer
miR-18a	Up-regulated	Oncomir/ suppressor	Thyroid gland anaplastic carcinoma, medulloblastoma, prostate cancer
miR-1246	Up-regulated	Oncomir/ suppressor	Pancreatic and colorectal adenocarcinoma
miR-200 family	Up-regulated/ down-regulated	Oncomir/ suppressor	Breast and colorectal cancer
miR-135a-3p, miR-200c, miR-216a and miR-340, miR-205	Up-regulated/ down-regulated	Suppressor	Ovarian cancer

The challenge of using biomarkers for disease diagnosis is to precisely identify which miRNAs can be reliably used as markers, or which can be used as therapeutic targets. As discussed earlier, conflicting observations about changes in miRNA levels in specific pathologies may arise. These differences may be attributed to various factors affecting the study such as sampling time, miRNA quantification methods, miRNA normalization parameters, and comorbidities (Zhou et al., 2018).

5.6.3 miRNAs as Therapeutic Agents

The miRNAs have been studied in recent years as potential therapeutics, as they can be used to target complementary mRNAs for degradation in a sequence-dependent manner. This redirection has been studied in the double membrane structures of extracellular vesicles (EVs), where a natural secretion process takes place and may be feasible for miRNA delivery. Different investigations have shown the promising therapeutic effects of EV-based nucleic acid delivery in cancer treatment (Meng et al., 2020).

Although the use of these molecules seems promising, the effects that these therapies may cause are still unknown and difficult to determine. The relationship between gene expression and miRNAs is complex. The silencing of a single miRNA would have the possible effect of modifying other genes, which can have diverse consequences, and this may not be entirely successful in multifactorial diseases. The use of miRNAs in the therapeutic field remains promising, even though it has not yet been possible to determine the adequate transport medium (e.g., viral or lipid vectors), which are now being extensively investigated (Specjalski and Jassem 2019).

Several pharmaceutical and biotechnology companies have been developing miRNA projects, mainly in two types of products: miRNA mimics and antagomiRs. The miRNA products are used to restore the concentration of a specific miRNA due to its suppression by the presence of a certain pathology. Conversely, antagomiRs are used to directly target the function of specific miRNAs, which are overexpressed and directly involved in a disease. To achieve integrated development in miRNA technologies, two main challenges must be addressed: first, the stability and specificity of the miRNA, and second, the specific delivery of miRNAs to the desired site of action. These problems can be addressed by passive applications involving the non-specific use of nanoparticles or actively using particles with the binding of a specific molecule that guarantees binding to the site of interest. These structural and delivery challenges, although constantly being addressed by new design strategies, still are major constraints hindering the development of miRNA therapeutics (Bonneau et al., 2019; Bajan and Hutvagner 2020).

5.7 RNA Editing

RNA editing is a cellular process in which mature RNAs are altered by enzymatic actions from the genomic sequence. Most RNA editing events in humans occur in the non-coding regions of mRNAs, such as introns and untranslated regions. RNA editing has several types of functional consequences, such as the regulation of different stages of miRNA biogenesis. This derived from the fact that miRNAs are generated from dsRNA intermediates. Furthermore, it has been described that RNA editing within miRNAs generates a change in miRNA targets. Similarly, RNA editing of miRNA binding sites within target transcripts alters miRNA targeting. To gain insight into the factors that influence the variability of RNA editing in different tissues, genetic and transcriptomic data are used for further analysis (Park et al., 2021).

In-depth study of RNA editing and its role in the regulation of biogenesis stages opens a very interesting door in biomedical research. An example is the editing of adenosine to Inosine (A to I) in the initial sequence of microRNAs. With this editing, the targetomes of microRNAs can be changed, and it has been shown that it can participate in pri-miRNAs, mature miRNAs, and targetome. For the case of ischemia, each editing results in a new miRNA with a unique targetome, leading to an increase in angiogenesis (van der Kwast et al., 2020).

To date, fewer than 160 miRNA editing sites, mainly miRNA A-to-I editing, have been identified through profiling of different samples (human, mouse, and *Drosophila*). Edition of miRNAs is prevalent in many tissue types in humans. Interestingly, miRNAs edited in neuronal and non-neuronal tissues in humans obtain new targets after editing; these are mainly associated with cognitive and organ developmental functions. This speaks of the impact of RNA editing on miRNA biology, and it is suggested that miRNA editing has recently acquired non-neuronal functions in humans (Li et al., 2018).

Several bioinformatics tools and pipelines have been designed for deeper analysis of RNA editing. These can be designed depending on the parameters used to filter and annotate reads, such as size exclusion minimum free energy to predict secondary structure, mismatches, and alignments with genomic libraries or available miRNA databases (e.g., miRbase or mirGeneDB). Some programs assess only the abundance of the edited miRNA forms, whereas others directly identify the targets of the edited miRNAs (e.g., SeqBuster, iMir and Prost!) and related genes. Targets are also identified by pathway enrichment as well as protein-protein interactions. This variability of protocols and pipelines dedicated to the identification of miRNA editing sites from sequencing data could strengthen the reliability of data obtained by different bioinformatics approaches but may also lead to differences and inconsistencies between studies (Lu et al., 2018; Correia de Sousa et al., 2019; Desvignes et al., 2019).

Last, the measurement of multiple omics profiles from the same single cell is important, as this helps to decipher the molecular regulation that exists at the intracellular level in disease development. This suggests that variability in microRNA expression alone may lead to non-genetic heterogeneity between cells (Wang et al., 2019).

References

Aghaei Zarch SM, Dehghan Tezerjani M, Talebi M, Vahidi Mehrjardi MY (2020) Molecular biomarkers in diabetes mellitus (DM). *Medical Journal of the Islamic Republic of Iran* 34: 28.

Alemán-Ávila I, Cadena-Sandoval D, Morales MJ, Ramírez-Bello J (2019) MicroRNA en enfermedades autoinmunes. *Gaceta Médica de México* 155(1): 63–71.

Bajan S, Hutvagner G (2020) RNA-based therapeutics: from antisense oligonucleotides to miRNAs. *Cells* 9(1): 137.

Balasubramanian S, Gunasekaran K, Sasidharan S, Jeyamanickavel Mathan V, Perumal E (2020) MicroRNAs and xenobiotic toxicity: An overview. *Toxicology Reports* 7: 583–595.

Bartel DP (2018) Metazoan microRNAs. *Cell* 173(1): 20–51.

Bautista-Sánchez D, Arriaga-Canon C, Pedroza-Torres A, De La Rosa-Velázquez IA, González-Barrios R, Contreras-Espinosa L, Montiel-Manríquez R, Castro-Hernández C, Fragoso-Ontiveros V, Álvarez-Gómez RM, Herrera LA (2020) The promising role of miR-21 as a cancer biomarker and its importance in RNA-based therapeutics. *Molecular Therapy. Nucleic Acids* 20: 409–420.

Beuzelin D, Kaeffer B (2018) Exosomes and miRNA-loaded biomimetic nanovehicles, a focus on their potentials preventing type-2 diabetes linked to metabolic syndrome. *Frontiers in Immunology* 9: 2711.

Bonneau E, Neveu B, Kostantin E, Tsongalis GJ, De Guire V (2019) How close are miRNAs from clinical practice? A perspective on the diagnostic and therapeutic market. *EJIFCC* 30(2): 114–127.

Cazzanelli P, Wuertz-Kozak K (2020) MicroRNAs in intervertebral disc degeneration, apoptosis, inflammation, and mechanobiology. *International Journal of Molecular Sciences* 21(10): 3601.

Chen B, Wu L, Cao T, Zheng H-M, He T (2020a) MiR-221/SIRT1/Nrf2 signal axis regulates high glucose induced apoptosis in human retinal microvascular endothelial cells. *BMC Ophthalmology* 20: 1–10.

Chen L, Heikkinen L, Wang C, Yang Y, Sun H, Wong G (2019a) Trends in the development of miRNA bioinformatics tools. *Briefings in Bioinformatics* 20(5): 1836–1852.

Chen P-S, Lin S-C, Tsai S-J (2020b) Complexity in regulating microRNA biogenesis in cancer. *Experimental Biology and Medicine* 245(5): 395–401.

Chen X, Xie D, Zhao Q, You Z-H (2019b) MicroRNAs and complex diseases: From experimental results to computational models. *Briefings in Bioinformatics* 20(2): 515–539.

Colhoun HM, Marcovecchio ML (2018) Biomarkers of diabetic kidney disease. *Diabetologia* 61(5): 996–1011.

Condrat CE, Thompson DC, Barbu MG, Bugnar OL, Boboc A, Cretoiu D, Suciu N, Cretoiu SM, Voinea SC (2020) miRNAs as biomarkers in disease: Latest findings regarding their role in diagnosis and prognosis. *Cells* 9(2): 276.

Correia de Sousa M, Gjorgjieva M, Dolicka D, Sobolewski C, Foti M (2019) Deciphering miRNAs' action through miRNA editing. *International Journal of Molecular Sciences* 20(24): 6246.

Daugaard I, Hansen TB (2017) Biogenesis and function of Ago-associated RNAs. *Trends in Genetics* 33(3): 208–219.

Denzler R, Stoffel M (2015) The long, the short, and the unstructured: A unifying model of miRNA biogenesis. *Molecular Cell* 60(1): 4–6.

Desvignes T, Batzel P, Sydes J, Eames BF, Postlethwait JH (2019) miRNA analysis with Prost! reveals evolutionary conservation of organ-enriched expression and post-transcriptional modifications in three-spined stickleback and zebrafish. *Scientific Reports* 9(1): 3913.

Dexheimer PJ, Cochella L (2020) MicroRNAs: From mechanism to organism. *Frontiers in Cell and Developmental Biology* 8: 409.

Doumatey AP, He WJ, Gaye A, Lei L, Zhou J, Gibbons GH, Adeyemo A, Rotimi CN (2018) Circulating MiR-374a-5p is a potential modulator of the inflammatory process in obesity. *Scientific Reports* 8(1): 7680.

Filipów S, Łaczmański Ł (2019) Blood circulating miRNAs as cancer biomarkers for diagnosis and surgical treatment response. *Frontiers in Genetics* 10: 169.

Gao Y, Patil S, Qian A (2020) The role of MicroRNAs in bone metabolism and disease. *International Journal of Molecular Sciences* 21(17): 6081.

Ghafouri-Fard S, Shoorei H, Taheri M (2020) miRNA profile in ovarian cancer. *Experimental and Molecular Pathology* 113: 104381.

Goodall GJ, Wickramasinghe VO (2021) RNA in cancer. *Nature Reviews Cancer* 21(1): 22–36.

He L, Zhu W, Chen Q, Yuan Y, Wang Y, Wang J, Wu X (2019) Ovarian cancer cell-secreted exosomal miR-205 promotes metastasis by inducing angiogenesis. *Theranostics* 9(26): 8206–8220.

Hess JF, Kohl TA, Kotrová M, Rönsch K, Paprotka T, Mohr V, Hutzenlaub T, Brüggemann M, Zengerle R, Niemann S, Paust N (2020) Library preparation for next generation sequencing: A review of automation strategies. *Biotechnology Advances* 41: 107537.

Huang Y, Yan Y, Xv W, Qian G, Li C, Zou H, Li Y (2018) A new insight into the roles of miRNAs in metabolic syndrome. *Bio Med Research International* 2018: 7372636.

Huang Z, Wu S, Kong F, Cai X, Ye B, Shan P, Huang W (2017) MicroRNA-21 protects against cardiac hypoxia/reoxygenation injury by inhibiting excessive autophagy in H9c2 cells via the Akt/mTOR pathway. *Journal of Cellular and Molecular Medicine* 21(3): 467–474.

Idda ML, Munk R, Abdelmohsen K, Gorospe M (2018) Noncoding RNAs in Alzheimer's disease. *Wiley Interdisciplinary Reviews. RNA* 9(2): e1463.

Iqbal MA, Arora S, Prakasam G, Calin GA, Syed MA (2019) MicroRNA in lung cancer: Role, mechanisms, pathways and therapeutic relevance. *Molecular Aspects of Medicine* 70: 3–20.

Jiang S, Yan W (2016) Current view of microRNA processing. *Signal Transduction Insights* 5: STI.S12317.

Kappel A, Keller A (2017) miRNA assays in the clinical laboratory: Workflow, detection technologies and automation aspects. *Clinical Chemistry and Laboratory Medicine (CCLM)* 55(5): 636–647.

Kempinska-Podhorodecka A, Blatkiewicz M, Wunsch E, Krupa L, Gutkowski K, Milkiewicz P, Milkiewicz M (2020) Oncomir microRNA-346 is upregulated in colons of patients with primary sclerosing cholangitis. *Clinical and Translational Gastroenterology* 11(1).

Kim K, Nguyen TD, Li S, Nguyen TA (2018) SRSF3 recruits DROSHA to the basal junction of primary microRNAs. *RNA* 24(7): 892–898.

Kobayashi H, Tomari Y (2016) RISC assembly: Coordination between small RNAs and Argonaute proteins. *Biochimica et Biophysica Acta (BBA) - Gene Regulatory Mechanisms* 1859(1): 71–81.

Kolenda T, Guglas K, Kopczyńska M, Sobocińska J, Teresiak A, Bliźniak R, Lamperska K (2020) 41—Good or not good: Role of miR-18a in cancer biology. *Reports of Practical Oncology & Radiotherapy* 25(5): 808–819.

Konovalova J, Gerasymchuk D, Parkkinen I, Chmielarz P, Domanskyi A (2019) Interplay between microRNAs and oxidative stress in neurodegenerative diseases. *International Journal of Molecular Sciences* 20(23): 6055.

Kozomara A, Birgaoanu M, Griffiths-Jones S (2019) miRBase: From microRNA sequences to function. *Nucleic Acids Research* 47(D1): D155–D162.

Li L, Song Y, Shi X, Liu J, Xiong S, Chen W, Fu Q, Huang Z, Gu N, Zhang R (2018) The landscape of miRNA editing in animals and its impact on miRNA biogenesis and targeting. *Genome Research* 28(1): 132–143.

Li Q, Xie J, Wang B, Li R, Bai J, Ding L, Gu R, Wang L, Xu B (2016) Overexpression of microRNA-99a attenuates cardiac hypertrophy. *PLoS ONE* 11(2): e0148480.

Lin S, Gregory RI (2015) MicroRNA biogenesis pathways in cancer. *Nature Reviews. Cancer* 15(6): 321–333.

Liu H, Liang C, Kollipara RK, Matsui M, Ke X, Jeong B-C, Wang Z, Yoo KS, Yadav GP, Kinch LN, Grishin NV, Nam Y, Corey DR, Kittler R, Liu Q (2016) HP1BP3, a chromatin retention factor for co-transcriptional microRNA processing. *Molecular Cell* 63(3): 420–432.

Liu X, Deng Y, Xu Y, Jin W, Li H (2018a) MicroRNA-223 protects neonatal rat cardiomyocytes and H9c2 cells from hypoxia-induced apoptosis and excessive autophagy via the Akt/mTOR pathway by targeting PARP-1. *Journal of Molecular and Cellular Cardiology* 118: 133–146.

Liu Y, Song J-W, Lin J-Y, Miao R, Zhong J-C (2020) Roles of microRNA-122 in cardiovascular fibrosis and related diseases. *Cardiovascular Toxicology* 20: 1–11.

Liu Z, Wang J, Cheng H, Ke X, Sun L, Zhang QC, Wang H-W (2018b) Cryo-EM structure of human dicer and its complexes with a pre-miRNA substrate. *Cell* 173(5): 1191–1203.e12.

Lu Y, Baras AS, Halushka MK (2018) miRge 2.0 for comprehensive analysis of microRNA sequencing data. *BMC Bioinformatics* 19(1): 275.

Matsuyama H, Suzuki HI (2019) Systems and synthetic microRNA biology: From biogenesis to disease pathogenesis. *International Journal of Molecular Sciences* 21(1): 132.

Meng W, He C, Hao Y, Wang L, Li L, Zhu G (2020) Prospects and challenges of extracellular vesicle-based drug delivery system: considering cell source. *Drug Delivery* 27(1): 585–598.

Mori MA, Ludwig RG, Garcia-Martin R, Brandão BB, Kahn CR (2019) Extracellular miRNAs: from biomarkers to mediators of physiology and disease. *Cell Metabolism* 30(4): 656–673.

Müller M, Fazi F, Ciaudo C (2020) Argonaute proteins: From structure to function in development and pathological cell fate determination. *Frontiers in Cell and Developmental Biology* 7: 360.

Nguyen Tuan A, Jo Myung H, Choi Y-G, Park J, Kwon SC, Hohng S, Kim VN, Woo J-S (2015) Functional anatomy of the human microprocessor. *Cell* 161(6): 1374–1387.

Nguyen Tuan A, Park J, Dang Thi L, Choi Y-G, Kim VN (2018) Microprocessor depends on hemin to recognize the apical loop of primary microRNA. *Nucleic Acids Research* 46(11): 5726–5736.

Nishi K, Takahashi T, Suzawa M, Miyakawa T, Nagasawa T, Ming Y, Tanokura M, Ui-Tei K (2015) Control of the localization and function of a miRNA silencing component TNRC6A by Argonaute protein. *Nucleic Acids Research* 43(20): 9856–9873.

O'Brien J, Hayder H, Zayed Y, Peng C (2018) Overview of microRNA biogenesis, mechanisms of actions, and circulation. *Frontiers in Endocrinology* 9: 402.

O'Neill RJ (2020) Seq'ing identity and function in a repeat-derived noncoding RNA world. Chromosome research: An international journal on the molecular, supramolecular and evolutionary aspects of chromosome *Biology* 28(1): 111–127.

Ouimet M, Ediriweera H, Afonso MS, Ramkhelawon B, Singaravelu R, Liao X, Bandler RC, Rahman K, Fisher EA, Rayner KJ, Pezacki JP, Tabas I, Moore KJ (2017) microRNA-33 regulates macrophage autophagy in atherosclerosis. *Arteriosclerosis, Thrombosis, and Vascular Biology* 37(6): 1058–1067.

Park E, Jiang Y, Hao L, Hui J, Xing Y (2021) Genetic variation and microRNA targeting of A-to-I RNA editing fine tune human tissue transcriptomes. *Genome Biology* 22(1): 77.

Park MS, Phan H-D, Busch F, Hinckley SH, Brackbill JA, Wysocki VH, Nakanishi K (2017) Human argonaute3 has slicer activity. *Nucleic Acids Research* 45(20): 11867–11877.

Parveen A, Mustafa SH, Yadav P, Kumar A (2019) Applications of machine learning in miRNA discovery and target prediction. *Current Genomics* 20(8): 537–544.

Perenthaler E, Yousefi S, Niggl E, Barakat TS (2019) Beyond the exome: the non-coding genome and enhancers in neurodevelopmental disorders and malformations of cortical development. *Frontiers in Cellular Neuroscience* 13: 352.

Plotnikova O, Baranova A, Skoblov M (2019) Comprehensive analysis of human microRNA–mRNA interactome. *Frontiers in Genetics* 10: 933.

Pong SK, Gullerova M (2018) Noncanonical functions of microRNA pathway enzymes–Drosha, DGCR8, dicer and ago proteins. *FEBS Letters* 592(17): 2973–2986.

Price NL, Singh AK, Rotllan N, Goedeke L, Wing A, Canfrán-Duque A, Diaz-Ruiz A, Araldi E, Baldán Á, Camporez J-P, Suárez Y, Rodeheffer MS, Shulman GI, de Cabo R, Fernández-Hernando C (2018) Genetic ablation of miR-33 increases food intake, enhances adipose tissue expansion, and promotes obesity and insulin resistance. *Cell Reports* 22(8): 2133–2145.

Riolo G, Cantara S, Marzocchi C, Ricci C (2020) miRNA targets: From prediction tools to experimental validation. *Methods and Protocols* 4(1).

Samidurai A, Kukreja RC, Das A (2018) Emerging role of mTOR signaling-related miRNAs in cardiovascular diseases. *Oxidative Medicine and Cellular Longevity* 2018: 6141902.

Sanjurjo MF, Calvo DdG, Robles SD, Dávalos A, Gutiérrez EI (2016) MicroRNA circulantes como reguladores de la respuesta molecular al ejercicio en personas sanas. *Archivos de medicina del deporte: revista de la Federación Española de Medicina del Deporte y de la Confederación Iberoamericana de Medicina del Deporte* 33(176): 394–403.

Schäfer M, Ciaudo C (2020) Prediction of the miRNA interactome–established methods and upcoming perspectives. *Computational and Structural Biotechnology Journal* 18: 548–557.

Singh T, Yadav S (2020) Role of microRNAs in neurodegeneration induced by environmental neurotoxicants and aging. *Ageing Research Reviews* 60: 101068.

Song R, Hu X-Q, Zhang L (2019) Mitochondrial miRNA in cardiovascular function and disease. *Cells* 8(12): 1475.

Specjalski K, Jassem E (2019) MicroRNAs: potential biomarkers and targets of therapy in allergic diseases? *Archivum Immunologiae et Therapiae Experimentalis* 67(4): 213–223.

Stavast CJ, Erkeland SJ (2019) The non-canonical aspects of microRNAs: Many roads to gene regulation. *Cells* 8(11): 1465.

Treiber T, Treiber N, Meister G (2019) Regulation of microRNA biogenesis and its crosstalk with other cellular pathways. *Nature Reviews Molecular Cell Biology* 20(1): 5–20.

van den Berg MMJ, Krauskopf J, Ramaekers JG, Kleinjans JCS, Prickaerts J, Briedé JJ (2020) Circulating microRNAs as potential biomarkers for psychiatric and neurodegenerative disorders. *Progress in Neurobiology* 185: 101732.

van der Kwast RVCT, Parma L, van der Bent ML, van Ingen E, Baganha F, Peters HAB, Goossens EAC, Simons KH, Palmen M, de Vries MR, Quax PHA, Nossent AY (2020) Adenosine-to-Inosine editing of vasoactive microRNAs alters their targetome and function in ischemia. *Molecular Therapy. Nucleic Acids* 21: 932–953.

Wang N, Zheng J, Chen Z, Liu Y, Dura B, Kwak M, Xavier-Ferrucio J, Lu Y-C, Zhang M, Roden C, Cheng J, Krause DS, Ding Y, Fan R, Lu J (2019) Single-cell microRNA-mRNA co-sequencing reveals non-genetic heterogeneity and mechanisms of microRNA regulation. *Nature Communications* 10(1): 95.

Wehbe N, Nasser SA, Pintus G, Badran A, Eid AH, Baydoun E (2019) MicroRNAs in cardiac hypertrophy. *International Journal of Molecular Sciences* 20(19): 4714.

Wojciechowska A, Osiak A, Kozar-Kamińska K (2017) MicroRNA in cardiovascular biology and disease. *Advances in Clinical and Experimental Medicine* 26(5): 868–874.

Wu Je, Yang J, Cho WC, Zheng Y (2020) Argonaute proteins: Structural features, functions and emerging roles. *Journal of Advanced Research* 24: 317–324.

Wu K-J (2020) The role of miRNA biogenesis and DDX17 in tumorigenesis and cancer stemness. *Biomedical Journal* 43(2): 107–114.

Xiang Y, Tian Q, Guan L, Niu S-S (2020) The dual role of miR-186 in cancers: Oncomir battling with tumor suppressor miRNA. *Frontiers in Oncology* 10: 233.

Xu Y-F, Hannafon BN, Khatri U, Gin A, Ding W-Q (2019) The origin of exosomal miR-1246 in human cancer cells. *RNA Biology* 16(6): 770–784.

Yao Z-Y, Chen W-B, Shao S-S, Ma S-Z, Yang C-B, Li M-Z, Zhao J-J, Gao L (2018) Role of exosome-associated microRNA in diagnostic and therapeutic applications to metabolic disorders. *Journal of Zhejiang University. Science. B* 19(3): 183–198.

Zhang T, Guo J, Gu J, Wang Z, Wang G, Li H, Wang J (2019) Identifying the key genes and microRNAs in colorectal cancer liver metastasis by bioinformatics analysis and *in vitro* experiments. *Oncology Reports* 41(1): 279–291.

Zhou S-S, Jin J-P, Wang J-Q, Zhang Z-G, Freedman JH, Zheng Y, Cai L (2018) miRNAs in cardiovascular diseases: Potential biomarkers, therapeutic targets and challenges. *Acta Pharmacologica Sinica* 39(7): 1073–1084.

6

Transcriptomics to Elucidate the Mechanisms of Pathogen–Human Interactions

Libia Zulema Rodriguez-Anaya
CONACYT–Instituto Tecnológico de Sonora, Mexico

Ángel Josué Félix-Sastré
Instituto Tecnológico de Sonora, Mexico

CONTENTS

DOI: 10.1201/9781003212416-7

6.1 Introduction

The human being interacts with microorganisms throughout his life. Some of these are beneficial, whereas others cause diseases. According to the World Health Organization (WHO), more than 20% of global deaths are caused by infectious diseases, where different types of microorganisms such as bacteria, fungi, viruses and protozoans are the pathogens responsible for these fatalities.

The pathogenesis of infectious diseases is determined by the interactions between the pathogen and the host immune response, also known as the interactome. As a result of these interactions, a cascade of events occurs in both organisms, triggering changes in their expression patterns in order to survive. These activations start at the moment that a pathogen enters the human body. The identification of these changes in the pathogen could lead to the discovery of virulence factors related to each disease. Through the last few years, the main tool used for the analysis of transcriptomes is RNA-seq, which enables highly accurate and sensitive transcriptional profiling. These studies allow the identification of both coding and non-coding RNAs, as well as their expression in a certain period of time. Traditionally, host–pathogen interaction studies with transcriptomics used to be limited to analyze the host and pathogen cells separately after the infection occurred. But, to truly understand all the mechanisms and interactions that occur during an infection, it would be mandatory to perform an analysis that allows the monitoring of gene expression in both pathogen and host cells. This is done by dual RNA-seq, a technique that discriminates expression profiles in silico through simultaneous read mapping with the respective reference genomes (Barquist et al., 2016). In this chapter, we will focus on different interactions between infectious agents and humans, along with the main diseases studied with transcriptomic profiling.

6.2 Human–Bacterial Interactions

In 2018, the WHO reported three infectious diseases inside the top ten causes of death worldwide. These diseases were lower respiratory tract infections (LRTIs), diarrheal diseases and tuberculosis. While the first two could be caused by both bacterial and/or viral agents, tuberculosis is exclusively caused by the bacterium *Mycobacterium tuberculosis*.

During an infection, multiple genes are expressed in both the pathogen and host cells. In order to grasp a better understanding of how bacteria, or any other pathogen cause infection, it is necessary to understand the expression and regulation of these genes (Saliba et al., 2017). The study of the complex interactions between bacterial pathogens, the commensal bacteria, and the host immune system is a rapidly emerging area of research. With transcriptomics and the rise of dual RNA-seq, it is now possible to have a better understanding of transcriptional and physiological changes during infection, which is necessary for the correct characterization of molecular phenotypes related to virulence. This has opened the path to uncover otherwise unknown virulence factors which could be undetected using standard infection assays, such as small non-coding RNA (sRNA) (Barquist et al., 2016). This section discusses the main transcriptomic responses in the host–pathogen interaction between host cells/immune system and the main causative agents of bacterial infections (Figure 6.1).

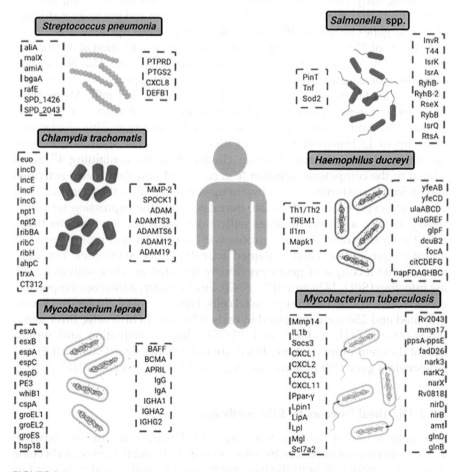

FIGURE 6.1
Main gene expression in host cells (blue) and bacteria (red) during infection. (Created with BioRender.com.)

6.2.1 Lower Respiratory Tract Infections (LRTIs) and Pneumonia

The LRTIs are the afflictions that cause the majority of deaths by infectious or communicable diseases in the world, and that globally are in the fourth cause of death, according to the WHO. The main causative agent for these infections is the opportunistic bacteria *Streptococcus pneumonia*. Its invasion on lower airways cause inflammatory and immune responses from the host, while it adheres to the epithelial cells (Hammerschmidt et al., 2007). Since the interaction between the pathogen and the lung epithelium activates numerous processes related to each other in both organisms, the study of transcriptional regulation during early and late infections is necessary to fully understand such interactions. With this objective, a dual RNA-seq study built a model replicating early LRTI with co-incubation of *S. pneumonia* and lung alveolar human epithelial cells. This resulted in the activation and posterior reduction of diverse gene clusters at different times post-infection (Aprianto et al., 2016).

A cluster of 33 *S. pneumonia* genes displayed gene activation at 30 minutes post-infection, followed by a reduced expression over time. This cluster was comprised of 20 genes related to carbohydrate transport and 13 encoders of adherence factors such as *aliA*, *malX*, *amiA*, *bgaA*, and *rafE*. At the same time, another gene cluster showed signs of down-regulation or repression, followed by up-regulation at later stages of the infection. This cluster was made up of 17 transporter genes of diverse substrates, such as iron, sugars, amino acids and ions. Another cluster of genes, containing 45 genes related to the competence regulon (e.g. *ccs4*, *cinA* and *dprA*) were activated at 60 minutes post-infection and were up-regulated until they reached their maximum expression levels at 240 minutes, when the expression remained steady. Finally, a cluster of genes with unknown functions was identified. It had a varied transcriptional profile with activation and repression at 30 minutes after infection and sustained activation after 60 minutes. This cluster included a couple of genes previously reported as being activated during infection (*SPD_1426* and *SPD_2043*). Furthermore, adherence-responsive genes were identified in the epithelial cells. From a pool of 272 genes, 19 were activated and 256 were repressed in early infection. In host cells, three genes (*ENSG00000237831*, *PTPRD*, and *PTGS2*) showed activation and repression at different points in time. Interleukin-8 encoder gene *CXCL8* (IL8) and β-defensin-1 gene *DEFB1* were also repressed (Aprianto et al., 2016).

6.2.2 Diarrheal Diseases by *Salmonella* spp.

Salmonella is one of the four key causes of diarrheal diseases worldwide. While most cases of infections by *Salmonella* are not mortal, they could become deadly depending on both the bacterium serotype and host-dependent factors. Approximately 2,600 serotypes of *Salmonella* have been identified to date, and most of them belong to the *S. enterica* species. The most common diseases caused by *Salmonella* spp. infection are gastroenteritis, bacteraemia and enteric fever, which are caused only by *S. typhi.* and *S. paratyphi* (Eng et al., 2015).

S. typhimurium is a very efficient pathogen when it comes to adaptation to new environments. Since transcriptional control is mediated by distinct regulatory proteins and sRNAs, there is a need to identify the different transcriptional signatures generated by each unique adaptation process. In 2013, 22 infection-relevant conditions were analyzed by differential RNA-seq. These environments include short environmental shocks such as peroxide shock and nitric shock; also, it gathers different ph and temperature conditions and nutrient limitation. The biggest changes in gene expression identified by the study occurred in the anaerobic, nitric oxide, osmotic and peroxide environmental shocks, where the expression of 15%–25% of chromosomal genes changed within 10 minutes. On average, 63% of genes were expressed in each environment, while 74% of genes were expressed on the total RNA pool. The changes in transcriptional signatures between conditions include the up-regulation of particular repair and detoxification genes, such as *ahpC*, *katG* and *oxyS* for peroxide exposure or *hmpA*, *norV* and *ytfE* for nitric oxide exposure. In addition, 3,838 transcriptional start sites (TSS) were identified (Kröger et al., 2013).

More recent transcriptomic studies on *Salmonella* infections were centered on the unknown functions of sRNA during invasion with dual RNA-seq. One study monitored the expression of sRNAs of *S. typhimurium* bacteria strain SL1344 located in fibroblast cells. It was found that intracellular bacteria induced and repressed different sRNAs at different times of the infection. Standing at the stationary phase in cell culture, the bacteria kept common expression trends for the sRNAs *DsrA*, *SsrS* and *SraL*, and increased expression of *RydC* and *Isrl*. Nonetheless, all were decreased upon entering the fibroblast cell and continued like that up to 24 hours post-infection (hpi). This decreased expression was also observed for *GlmZ*, *SroC* and *IsrH*- after intracellular invasion. At 2 hpi, the sRNAs *InvR*, T44 and *IsrK* showed up-regulation with constant decay after that, which indicates a role for gene targeting during early infection. At late infection times, an accumulation of *IsrA*, *RyhB-1*, *RyhB-2*, *RseX*, *RybB* and *IsrQ* sRNAs was observed, and these reached their maximum levels between 6 and 24 hours after initial infection. Also, one study focused on *RyhB-2* revealed that it had down-regulation activity targeting the *YeaQ* gene (Ortega et al., 2012). Another study related to non-coding RNA functions identified a Phop-activated sRNA named *PinT*. This RNA showed an increase up to 100-fold during infection and general up-regulation in several cell types, ranging from murine bone marrow macrophages to human THP-1 macrophages. This 80 bp sRNA is located in a locus that also encodes for a co-activator for invasion genes designated as *RtsA*. It has been defined to play a support role in virulence due to its sequence conservation and strong induction inside of host cells (Barquist et al., 2016).

The encounter between cells of the immune system and any invading bacteria initiate a series of interactions that develop the course of infection. The variation between cells and the different environmental factors causes different expression patterns in both the bacteria and the immune cell. In a study with a combination of single-cell RNA-seq (scRNA-seq) and fluorescent markers, it was determined that the variations between infected host cells are caused by the heterogeneous activity of the invading bacteria. The study revealed specific genetic pathways that exhibit considerable variation between infected host cells, such as the Type I IFN response, in which genes were induced in one-third of infected macrophages in their experiment at 4 hpi. It also presented a bimodal gene expression at the 8-hour mark, a time at which such genes were also induced in uninfected macrophages. This variation was proved to be dependent on the expression of the bacterial transcription factor PhoP (*phoP* gene) and the cognate sensor kinase PhoQ (*phoQ* gene). Other pathways, such as cyctokine-cyctokine receptor interactions, Rig-I receptor signaling and Toll-like receptor signaling, showed high variance between cells 8 hours after initial exposure. With this, it was deduced that genes induced by intracellular bacterial signals showed higher variation than those induced by extracellular exposure cues in infected cells (Avraham et al., 2015).

Another study, using a then-novel single-cell-RNA sequencing method that allowed getting both host and pathogen transcriptomes, defined as scDual-seq, detected three different linear progressions for different infected macrophages: infected cells with an "induced" response, infected cells whose transcriptomes resembled those of unexposed cells, named as "partially induced", and unexposed cells in general. Even though the transcriptional patterns of the partially induced macrophages are similar to the unexposed immune response genes such as *Tnf* and *Sod2*, those patterns are differentially expressed between them (Avital et al., 2017).

6.2.3 Tuberculosis

The bacterium *Mycobacterium tuberculosis* is the pathogen responsible for this disease and is mainly located in the lungs. Unfortunately, according to the WHO, over 10 million people are infected with *M. tuberculosis* and at least 1.5 million die from it each year. Despite being a disease that is preventable and curable, it still is the leading cause of death for those infected with HIV and is a major concern due to the existence of antimicrobial resistance strains. The main issue is that as an intracellular pathogen, *M. tuberculosis* does not possess common virulence factors, since it can persist within a host body in a long-term latency or dormant state, without causing damage, unless the host immune system is compromised (Chai et al., 2018). For this reason, the application of transcriptomics to identify more unique virulence factors is essential to understanding more of this particular pathogen and the host immune response to it.

A dual RNA-seq analysis of *M. tuberculosis* infected macrophages *in vivo* revealed up-regulated genes related to infection. In this particular analysis by Pisu et al. (2020), this happened only in transcripts with a Log FC value > 1 and an adjusted p value < 0.05, that were up-regulated only on alveolar and interstitial macrophages. This analysis identified 180 genes that fulfilled these conditions, and such genes were found to be related to several pathways, specifically: synthesis of cell wall phthiocerol dimycocerosates (*Rv2943, mmpl7, ppsA-ppsE* and *fadD26*); nitrate/nitrite detoxification (*narK3, narK2, narX, Rv0818, nirD* and *nirB*); and ammonia uptake (*amt, glnD* and *glnB*). In case of transcriptional responses to the infection by macrophages, these showed the up-regulation of triglyceride and cholesterol metabolism (*Mgl, Lpl, LipA, Lpin1* and *Ppar-γ*).

A global gene expression by the host in response to *M. tuberculosis* has been identified through the use of RNA-seq analysis applied to macrophages infected with the virulent *M. tuberculosis* strain H37Rv and the avirulent strain H37Ra, in order to contrast the immune response to actual virulence-related transcripts. This analysis revealed 750 DE genes found in response to both strains in contrast to a control. Regardless of the strain infecting the macrophage, these showed overexpression of genes related to the TNF signaling pathway (*Mmp14, IL1b, Socs3, Cxcl1, Cxcl2, Cxcl3* and *Cxcl1*). Between both strains, only 85 genes were found to be DE; among these, the main gene found to be related to *M. tuberculosis* survival was a solute carrier gene (*Slc7a2*), which was 17.4 times overexpressed in macrophages infected with the avirulent strain in comparison to the virulent one. By silencing this gene through selected siRNAs targeting it, it was discovered that in absence of it, the avirulent strain intracellular survival increased significantly, while the virulent strain showed no change in survivability, concluding that *M. tuberculosis* virulent strains silence this gene in order to survive within the host macrophages (Lee et al., 2019).

In case of diseases similar to tuberculosis, some research has been done on *Yersinia pseudotuberculosis* to identify host and pathogen interactions. Applying tissue dual RNA-seq on *Y. pseudotuberculosis* cultivated in lymphoid tissue showed a significant alteration on the transcription profile on host cells post-infection. From a total of 19,993 profiled transcripts, 1,336 were significantly altered. Within these, 448 transcripts were induced, and 888 were repressed in consequence of the infection. The highly induced genes identified here were related to inflammatory responses, coagulative activities, acute-phase responses and transition metal ion sequestration. The transcriptional pattern showed a relation between inflammatory responses like inflammatory cytokines (*IL-6, IL17F, IL-1α, IL-1β* and *IFN-γ*) and chemokine receptors (*CXCL1, CXCL2, CXCL3, CXCL5, CXCL10* and *CCL2*) with neutrophil-specific transcripts (*prokineticin 2, Schlafen 4, Fpr1*, etc.). This demonstrated that the host immune response was directly altered by infiltrated neutrophils. The genes related to the acute-phase response were also enriched after the infection. This gene group was mainly composed of serum amyloids such as *P*, *A2*, *A3* and *A4*, which contribute to intestinal immunity and that are themselves up-regulated by proinflammatory signals. These contribute in the modulation of both adaptive and innate immune responses. In this study, transcripts of metalloproteases *MMP3* and *MMP8*, and chitinase-like proteins such as *CHIL1* and *CHIL3*, were significantly enriched. On the other hand, the most enriched messengers of the host cells were the coagulative factors *FII*, *VII*, *X* and *XIIIa1*, the subunits α-, β- and γ of the reactant fibrinogen and plasminogen *Plg* (Nuss et al., 2017). Also, especially high enrichment of factors *VII*, *X* and fibrinogen subunits indicates that the *Yersinia* infection triggers the activation of the tissue factor-dependent extrinsic coagulation cascade (Nuss et al., 2017).

Metal ion sequestration factors were found to be among the strongest induced by the infection, with genes such as haptoglobinand lipocalin 2 (*LCN2*), that were induced by more than 100-fold post-infection and others like lactotransferrin and hemopexin that were 7- and 14-fold enriched only. Other unrelated genes like *Irg1* and *Hdc* were also highly induced by the pathogen. Finally, in the case of the 888 down-regulated genes, several were monooxygenases (cytochrome P450 enzymes, solute carriers and UDP-glucuronosyltransferase). As for the depleted transcripts, most were related to transport and hydrolytic genes of saccharides (*Slc5a4a*, *Slc5a11*, *Slc5a4b*, *Slc5a1*, lactase, trehalase and sucrose isomaltase) (Nuss et al., 2017).

In the transcriptional host-responsive profile of *Yersinia*, it was found that the thermal upshifts upon entering the host triggers the expression of the pYV virulence factor. Furthermore, the expression of such a transcript was further increased during colonization, while down-regulators of this gene such as *copB* were significantly reduced. Also, it was recorded the host-induced expression of stress response genes, general stress resistance genes (e.g., *dnaK*, *grpE*, *groEL/ES*, *rpoS*), other genes related to nitric oxide detoxification like nitric oxide dioxygenase: *hmp* and genes related to iron uptake systems (*fcuA*, *rhbC*, *sfuB*, *tonB*, etc.) (Nuss et al., 2017).

6.2.4 Chancroid

Haemophilus ducreyi is a gram-negative bacteria and the causative agent of a genital ulcerative disease known as chancroid. It is the major cause of this genital disease in both Africa and Asia. The presence of this disease facilitates the transmission of the human immunodeficiency virus type 1 (Ronald & Plummer, 1985). In a study to define the *H. ducreyi*–human interactome, dual RNA-seq was applied by the inoculation of the bacteria in the arm of human volunteers. This revealed that *in vivo*, *H. ducreyi* had 218 DE genes in comparison to *in vitro*. Of these, 81 monocistronic and 80 polycistronic operons were found, and 113 genes were up-regulated and 105 were down-regulated. The up-regulated group included genes related to pathways such as transporters, replication and repair, carbohydrate metabolism, signal transduction, secretion and translation. Other up-regulated genes were those related to manganese metabolism (*yfeAB*), iron metabolism (*yfeCD*), aldarate metabolism (*ulaABCD* and *ulaGREF*) and glycerol metabolism (*glpF*) alongside specific genes involved in anaerobic respiration (*dcuB2* and *focA*), citrate metabolism (*citCDEFG*) and periplasmic nitrate reductase genes (*napFDAGHBC*). Genes that were down-regulated belong to pathways like ribosome biogenesis, chaperones and nucleotide metabolism. On the other hand, host human cell gene expression revealed 1,873 up-regulated and 1,007 down-regulated genes in comparison to non-infected cells. The up-regulated pathways included T cell activation genes (Th1/Th2 and T helper pathways) and innate response (TREM1, NK cells, signaling, cross talk, and phagosome formation pathways). The two encoding regulators with the most significant expression change were the down-regulated Il1rn and Mapk1 (Griesenauer et al., 2019).

6.2.5 *Chlamydia trachomatis* Infection

Infections caused by *Chlamydia trachomatis* are sexually transmitted and cause a great deal of reproductive afflictions such as infertility, pelvic inflammatory disease and ectopic pregnancy. Unfortunately, the mechanisms of tissue damage induced by this pathogen are unknown (Zanto et al., 2010).

Dual RNA-seq was applied to obtain transcriptomes for both *C. trachomatis* (serovar E EBs) and infected epithelial cells (HEp-2) at 1 and 24 hours after initial infection. At early infection time (1 hour), approximately 153 genes (approximately 17% of the total genetic pool) were highly expressed, and 56 of these were only up-regulated at this stage of early infection, including known genes such as the master regulator *euo*, inclusion proteins *incD*, *incE*, *incF* and *incG*, translocases *npt1* and *npt2*, and riboflavin biosynthetic enzymes *ribBA*, *ribC* and *ribH*, proteins related to iron acquisition in many bacterial pathogens (Crossley et al., 2007; Humphrys et al., 2013). Other novel genes to this stage were found, of which 25 are hypothetical proteins with gene ontology (GO) terms like transmembrane transport, carbohydrate metabolic process and biosynthetic process. On the other hand, at 24 hpi, *C. trachomatis* showed up-regulation of 112 genes that were only highly expressed at this time. These included previously undescribed proteins like peroxidase *ahpC*, thioredoxin *trxA* and ferredoxin CT312. Since host cells generate reactive oxygen species (ROS) in response to chlamydia infections, these genes might be the pathogen response to ROS (Humphrys et al., 2013).

Meanwhile, the analysis of host HEp-2 cells identified 82 DE genes at 1 and 24 hours, and 4 DE at 24 hours only. The GO analysis revealed that these genes were mainly related to terms such as inflammatory response, anti-apoptosis and immune response. Examples of up-regulated genes found at both early and late infections were: metalloproteinase *MMP-2*, chimeric proteoglycan *Testican-1/SPOCK1* (*MMP-2* regulator), and *ADAM/ADAMT* proteins *ADAMTS3*, *ADAMTS6*, *ADAM12* and *ADAM19*. Several other proteins of the basement membrane components, such as members of the collagen superfamily, latimin heterotrimers and nidogen, were also up-regulated. All these proteins were related to the process of chlamydial scarring (Humphrys et al., 2013).

6.2.6 Leprosy

Leprosy is an infection caused by *Mycobacterium leprae* that targets mainly nerves and skin. It enters the organism through the respiratory system, migrates towards the neural tissue and then enters the Schwann cells. The bacteria itself has low virulence since the majority of infected individuals do not develop the disease. The success of the infection depends entirely on the person's resistance or immune response (Joshi & Srivastava, 2009).

To study the host–pathogen interaction of this bacteria *in situ*, dual RNA-seq was applied on 24 leprosy skin biopsy specimens. With the application of a gene signature-based analysis, a significant correlation between the abundance of *M. leprae* and the gene expression signature of IFN-β was found. Despite this, it had no relation with the IFN-γ signature. Also, the abundance of *M. leprae* did not correlate with a macrophage gene expression signature. Furthermore, most of the sequenced transcriptome was composed by structural sRNAs. Of all the mRNAs present in the transcriptome, the most abundant were virulence-related proteins, such as *esxA* and *esxB*, that compose the secretion system ESX1; the operonic transcripts *espA*, *espC*, *espD* and *PE3* that comprise the ESX1-associated proteins; and the master transcriptional regulator whiB1 and stress protein transcripts such as *cspA*, *groEL2*, *groES* and *hsp18*. On the other hand, while looking for a correlation between *M. leprae* and host response expression, the most expressed genes concerning *M. leprae* infection were those related to the humoral immune response and overall immunoglobulin genes including all immunoglobulin heavy chain constant genes (IGH). A study done with ingenuity pathway analysis (IPA) showed that the most enriched pathway was related to the communication between innate and adaptive immune cells. This contains genes known as the B-cell-activating factor (*BAFF*) and a proliferation-inducing ligand (*APRIL*) that activates the B-cell maturation antigen (*BCMA*) that leads to the production of immunoglobulin (*IgG*) and A (*IgA*). Further, genes *IGHA1*, *IGHA2* and *IGHG2* were inversely correlated with *M. leprae* mRNA/rRNA ratio. It was also identified that *M. leprae* genes *aac*, *clpB*, *hspR*, *groEL1*, *groEL2* and *ML0493* correlated with the expression of host cells *BCMA*, *IGHA1*, *IGHA2* and *IGHG2* genes (Montoya et al., 2019).

6.3 Human–Fungal Interaction

Over the last few years, fungal infections have increased astronomically, infecting ever more people in nosocomial conditions, even though fungal

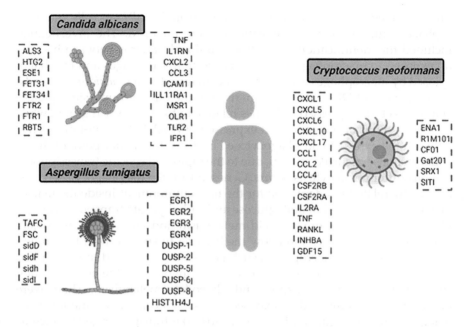

FIGURE 6.2

Main gene expression in host cells (blue) and fungi (red) during infection. (Created with BioRender.com.)

infections range from mild skin lesions that do not present any threat, to mortal invasive infections such as fungemia. Fungal infections mainly affect immunocompromised patients. Infections such as those caused by *Candida albicans* can inflict serious diseases, but in healthy individuals, such a yeast would stay as a common resident of skin, the urogenital tract and gastrointestinal system. Invasive fungal infections can cause mortality rates between 38% and 63%, and they kill around 1.5 million people worldwide. The most common fungal pathogens belong to the genera *Candida*, *Cryptococcus* and *Aspergillus* (Hovhannisyan & Gabaldón, 2018; Muñoz et al., 2018). This section will describe the main genes related to the host–pathogen interaction of the most common fungal infections (Figure 6.2).

6.3.1 Candidiasis

Candida albicans is a fungus that is part of the human microbiome. Nevertheless, in individuals with compromised or deficient immune systems, it could become a devastating and even mortal pathogen. Similar to other pathogens, from a direct interaction between macrophages and the fungus, a variety of subpopulations with heterogeneous infection outcomes have been reported. For this reason, it is necessary to uncover the full relation or interactome between the immune host cells and *C. albicans*. In a study that examined the expression of differently treated *C. albicans* cells and macrophages, different expression patterns were unveiled. The investigation included the identification of the transcriptional patterns between bacteria (a) exposed to macrophages, (b) unexposed to macrophages and (c) phagocytosed by macrophages across time. The phagocytosed *C. albicans* showed a large quantity of DE genes in comparison to the other groups, since phagocytosed bacteria expressed 732 genes relative to the unexposed group, while the exposed but unphagocytosed expressed only 82, both at all-time points. Between the exposed and phagocytosed bacteria, the higher point of differential response occurred at 1 hpi, due to the rapid and specific transcriptional response after phagocytosis occurs. Genes induced in the phagocytosed cells are mostly related to adaptation for the new environment inside the macrophage, genes involved in both glucose and carbohydrate transports, carboxylic acid metabolism, organic acid metabolism and fatty acids catabolism. These changes in expression allow the bacteria to survive with the limited nutrients inside the macrophage; such genes were repressed in unengulfed bacteria. There is also up-regulation in genes involved in the beta-oxidation cycle, transmembrane transport and glyoxylate metabolism. These were highly induced compared to exposed *C. albicans* cells. Even though various genes are up-regulated in both groups, like transporter genes, such as high-affinity glucose transporters, amino acid permeases and oligopeptide transporters, this suggests that these genes are not related to the phagocytosis. Alike, genes related to pathogenesis and formation of hyphae, such as core filamentous response genes like *ALS3*, *ORF19.2457*, *HTG2* and *ESE1*, were up-regulated from the initial infection until 4 hours after infection. This result suggests that these are genes expressed in response to macrophage presence. Other genes related to oxidation reduction, such as transcription factors, dehydrogenases and genes related to mitochondrial respiratory response, were also up-regulated during phagocytosis, but not as much as those mentioned earlier. Also, a number of genes in engulfed bacteria, such as genes related to ribosomal proteins, transcription factors, chaperones and genes related to translation machinery, were repressed at early time points but regained expression after 4 hpi. Likewise, genes related to ergosterol biosynthesis, cell growth and cell wall synthesis were also repressed. Some not strongly repressed genes in phagocytosed bacteria were those related to nucleoside metabolism (Muñoz et al., 2018).

Regarding more expressed virulence factors within *Candida albicans*, applying RNA-seq during treatment using filamentation inhibitors resulted in virulence-related genes showing down-regulation: multicopper oxidases *FET31* and *FET34*, iron permeases *FTR2* and *FTR1*, and cell wall protein *RBT5* (Romo et al., 2019).

Analysis on the transcriptional response from macrophages was done in parallel with *C. albicans*. In this case, the transcriptional profiles of both exposed and infected macrophages were closely related, since both types showed up-regulated genes related to phagocytosis, migration and triggering of innate immune responses. These included the induction of pathways such as NF-kB signaling, *IL-8*, *IL-6*, *Fcy* receptor-mediated phagocytosis, pattern recognition receptors, *RhoA*, *ILK*, Leukocite Extravasation signaling and production of both nitric oxide and *ROS*. It was also observed that the up-regulation of some sets of genes were kept constant even after the phagocytosis of *C. albicans*. Genes related to pro-inflammatory cytokine production (*Tnf*, *Il1rn*, *Ccl3*, *Cxcl2* and *Cxcl14*) and fungal recognition via transmembrane receptors (*Icam1*, *Il11ra1*, *Msr1*, *Olr1*, *Tlr2* and *Ifr1*) were induced after the first hour and remained stable through 2 and 4 hpi. Another set of up-regulated genes in both exposed and infected macrophages peaked expression at 2 hpi and were associated with pathways such as pathogen recognition, engulfment activation, opsonization and genes *Dectin-1*, lectin-like receptors (*Lgals1* and *Lgals3*) and transmembrane receptors (*Cd74*, *Trem2*, *Gm14548*, *Itgb2*, *Cd36*, *Fcer1g*, *Ifnar2* and *Igsf66*). Multiple complement proteins had higher induction in only exposed macrophages: complement factor properdin (*Cfp*), complement receptor proteins (*C3ar1* and *C5ar1*) and extracellular complement proteins (*C1qa*, *C1qc* and *C1qb*). These results indicate the importance of bacterial recognition and activation of engulfment of macrophages. Subpopulations of macrophages infected with *C. albicans* showed high induction of pro-inflammatory cytokines *Ccl3*, *Tnf*, *Cxcl2*, *Cxcl4*, *Il1rn* and transcription regulators *Irf8*, *Nfkbia* and *Cebpb*. The expression of these genes seem to be necessary for clearance of live *C. albicans* after phagocytosis.

After 4 hpi, a shift in gene expression was observed, with either high repression or high induction of genes. The group of repressed genes included cytokines *Il1a* and *Irf4*, intracellular Toll-like receptors *Tlr9* and *Tlr5*, and interleukin receptors *Il17ra*, *Irf4* and *Il1r1*. On the other hand, the group of induced genes gathered transcriptional regulators related to inflammation and apoptosis (*Fos*, *Cebpd*, *Bcl2*, *Card9*, *Nfkbie*, *Irf8*, *Irf9* and *Irf22bp1*) (Muñoz et al., 2018).

6.3.2 Aspergillosis

Aspergillus fumigatus is the main pathogenic species of *Aspergillus* and the primary causative agent for infections of the genus. As for most fungal infections, this pathogen can cause diverse kinds of diseases or ailments depending on the host immune system status. People with altered lung function can develop allergic bronchopulmonary aspergillosis, repeated exposure to conidia may lead to the formation of noninvasive aspergillomas in lung cavities and immunocompromised individuals are vulnerable to invasive aspergillosis, the most dangerous disease caused by *Aspergillus*. Lung epithelial cells function as the first defense barrier against pathogens, both physical and immunological, since they release inflammatory factors, macrophages, neutrophils and also signal lymphocytes (Dagenais & Keller, 2009).

Standard RNA-seq has been used to assess the transcriptional response of cultured alveolar epithelial cells after infection of *A. fumigatus*. From a total of 19,118 genes, 459 were DE. In comparison with uninfected cells, 302 genes were up-regulated and 157 were down-regulated in epithelial cells infected with the fungus. With GO enrichment analysis, it was found that the up-regulated genes were mainly associated with terms such as immune response, positive regulation of macromolecule metabolism, cell activation chemotaxis, regulation of phosphorylation, response to extracellular or endogenous stimulus and defense response, inflammatory response, response to bacterium and positive regulation of transcription. Within these genes, early grown response 4 (*EGR4*) had the maximum fold change. Other genes of the EGR family were also up-regulated (*EGR1, EGR2* and *EGR3*) along with five dual-specificity phosphatases (DUSPs) genes (*DUSP-1, DUSP-2, DUSP-5, DUSP-6* and *DUSP-8*. On the other hand, down-regulated genes included genes associated with terms such as vasculature development, skeletal system development, immune system development, ion transport and hemopoiesis (Chen et al., 2015). The down-regulated gene with the maximum fold change was an encoder for a core component of nucleosome, histone *HIST1H4J*. Infection with *A. fumigatus* induces cytoskeleton rearrangement of lung epithelial cells (Jia et al., 2014). This study showed that lung epithelial cells up-regulate the expression of genes that play an important role in remodeling of cytoskeleton, morphological change and regulation of actin polymerization, as a response against *A. fumigatus* infection (Chen et al., 2015).

In case of *A. fumigatus* response to a host immune response, an analysis of the transcriptomes of two *A. fumigatus* strains (*Af293* and *CEA10*) during infection revealed two main gene groups shared between both strains and showed up-regulation at some point in time during infection: siderophores *TAFC* and *FSC*, and ligases *sidD, sidF, sidH* and *sidI* (Watkins et al., 2018).

6.3.3 Cryptococcal Meningitis

Cryptococcus neoformans is one of the most frequent causative agents of fungal diseases such as Cryptococcal meningitis. However, it is not a well-adapted pathogen in humans, since it causes serious infections only on really immunocompromised individuals. Nevertheless, this fungus rarely interacts with human hosts. This means that most of its virulent phenotypes are related to a rapid and effective capacity to adapt to varying environments and conditions in both the external environment and inside a human host (Alspaugh, 2015).

RNA-seq has been used to analyze global gene expression, DE and genetic diversity of clinical strains of *C. neoformans* under exposure to cerebrospinal fluid (CSF) *in vivo, ex vivo* and *in vitro* (cultured with yeast extract) to identify virulence factors, genes and pathways related to its survival in the human central nervous system and further infection. Cells cultured *ex vivo* showed 129 DE genes against *in vitro, in vivo* revealed 45 DE genes against *in vitro* and *in vivo* showed 256 DE genes against *ex vivo*. Transcriptional profiles of *in vivo* and *in vitro* cultures were more similar between them than with *ex vivo* cultures. Functional categories of the transcripts via GO analysis revealed that C. *neoformans* cells cultured *in vivo* up-regulated genes related to cell metabolism processes like gene expression, structural constituents of the ribosome and biosynthetic processes. This showed that the pathogen was more biosynthetically active in exposure to host inflammatory cells. On the other hand, genes up-regulated in both cultures with CSF are related to survival and fitness of C. *neoformans* within the human subarachnoid space, since there were multiple genes up-regulated in these conditions that have been reported as putative virulence or fitness genes, such as *ENA1*, *RIM101* and *CF01*. However, other genes could not be functionally annotated, and a couple of them were identified as putative target genes of the virulence regulator *Gat201*. Also, specific transcriptional responses of the pathogen within the human body were found. Six genes were significantly up-regulated *in vivo* compared to the other conditions. Among these, gene *SRX1* has been reported to play an important role in resistance against oxidative stress, and gene *SIT1* is essential for melanin formation and growth under low-iron conditions (Chen et al., 2014).

Regarding immune responses in presence of this fungus, using mouse and monkey models, Li et al. (2019) applied KEGG analysis, identifying 40 different pathways. Cytokine-cytokine receptor interaction, tumor necrosis factor signaling, chemokine signaling and Toll-like receptor pathway. During invasion of lung tissue in both animal models, several induced genes were identified in common: chemokine transcripts (*CXCL1*, *CXCL6*, *CXCL5*, *CXCL17*, *CXCL10*, *CCL4*, *CCL2*, *CCL1*, etc.), chemokine receptors (*CSF2RB*, *CSF2RA* and *IL2RA*), TNF family transcripts (*TNF* and *RANKL*) and TGF-ß family genes (*INHBA* and *GDF15*). This represents a global immune response to the infection caused by the pathogen in these species and potentially also in humans.

6.4 Human–Viral Interactions

Viruses are extremely different from the rest of pathogens known to humans. They have a very specific route of action and require the inner cellular machinery to replicate and cause disease. A viral disease requires the implantation of the virus at a particular portal of entry, local replication on target cells, transportation or spread to specific tissues or organs and spread to the viral shedding site. Also, different factors exist that inhibit or enhance the infection, since viruses depend on their accessibility to a target tissue, host cell susceptibility to allow viral replication and the virus's own susceptibility

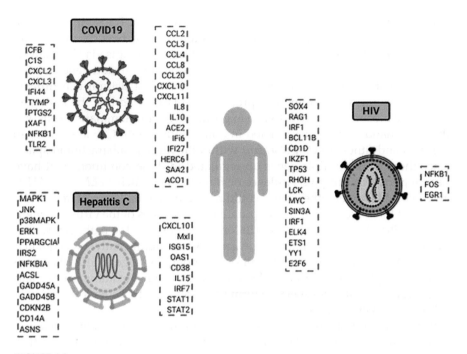

FIGURE 6.3
Main gene expression in host cells (blue) and viruses (red) during infection. (Created with BioRender.com.)

to the host immune system (Baron, 1996). Since viral infections are so deeply related to transcriptional mechanisms, it is critical to apply transcriptomics techniques to unveil all interactions between the virus and the host. Many of the diseases caused by viruses are devastating, like HIV, HCV and COVID-19. In this section, we will show the most expressed genes related to the pathways taken during infection for both host cells and viruses (Figure 6.3).

6.4.1 Human Immunodeficiency Virus (HIV)

Globally, 37.9 million people live infected with HIV, while only 23.3 million receive antiretroviral treatment. HIV mainly targets CD4 T lymphocytes and infects all cells with CD4 or chemokine receptors, from resting CD4 T cells to macrophages. The main effect of HIV is the dysfunction of the immune system due to the progressive depletion of CD4 T cells, since they are the target cells directly eliminated by the infection. Furthermore, this affliction also produces preferential loss in T-helper-17 cells, causing even greater loss of crucial defense responses against pathogens. Nowadays, the use of anti-retroviral therapy has managed to suppress viral infections from progressive fatal illnesses to a manageable chronic disease (Maartens et al., 2014).

Viral infections such as HIV directly modify the transcription of the host cell, inducing an alarming expansion of viral expression and repression of host genes. In a study using a T lymphoblast model of HIV infection, RNA-seq was applied to identify transcriptional profiles at early times of infection and the impact of viral mRNA on infected cells. At 12 hpi, viral mRNA could be detected at appreciable levels, but most cells seemed to be intact, with just some expressing the viral antigen p24 Gag. At 24 hpi, viral mRNA was highly expressed, and almost all cells expressed the antigen. This time was the peak of viral RNA replication. Both times (12 and 24 hpi) were selected as timestamps for analysis. At 12 hpi, viral reads constituted 18% of total reads, while at 24 hpi they represented 38% of the reads. Then, DE host genes were identified; a total of 106 genes were DE at 12 hpi, and by 24 hpi that number raised to 5,006 DE genes. In total, 97 of the 106 DE genes at 12 hpi kept the same expression level, even after 24 hpi. The most overexpressed DE genes found at 12 hpi were those related to T cell differentiation, followed closely by protein kinase regulation, biosynthetic process regulation, protein kinase inhibitor activity and caspase activity regulation. On the other hand, at 24 hpi the most overexpressed genes were those annotated as ribonucleo-protein complex biogenesis related, and those related with nucleotide bind-ing, cell cycle, RNA processing, DNA metabolic process, Macromolecular complex organization, Chromosome organization, Lymphocyte differentia-tion, Posttranscriptional regulation of expression and RNA transport. At 12 hpi, seven DE genes related to T cell activation (*SOX4, RAG1, IRF1, BCL11B, CD1D, EGR1* and *IKZF1*) were found, of which six were down-regulated indicating suppression of T cell activation. Five of these encoded transcrip-tional factors (*SOX4, IKZF1, EGR1, IRF1* and *BCL11B*), are relevant to either immune response and inhibition of HIV expression. The CD1D product is directed by the HIV-1 Nef for internalization and degradation, where it should present antigens on the cell surface, so its down-regulation reduces surface expression and facilitates immune evasion by pathogens. Another product is BCL11B which directly binds to HIV long terminal repeat; there-fore, its repression facilitates HIV replication. It was also found that by 24 hpi the infection successfully reprograms the transcriptional profile of the cells, since DE genes associated to fundamental cellular functions and a wide range of major biological processes are overrepresented, such as DNA metabolism, transcription, cell cycle control, protein degradation and mRNA processing. All of these specific genes were detected to be strongly down-regulated at that time: *CD3D, CD3Q, RHOH, TP53, LCK, CD1D, CD28, TREML2, TNFRSF4* and *CXCR4* (S. T. Chang et al., 2011). On another note, the application of Total RNA-seq potentially allows coverage of all fractions of the transcriptome and unravels the early and late host cell responses to HIV infection. Total RNA-seq and DE analysis of a CD4+ T cell line infected with intact and UV-inactivated HIV-1 revealed 11,094 DE genes following HIV infection. Evaluation of the transcriptional profile of T cells infected with inactivated HIV revealed that nonreplicating HIV virions induced expression changes strikingly similar to those infected with unaltered HIV virions. This indicates that the host cell response to early HIV infection is independent of viral replication, with the difference residing in the magnitude of expression changes, which was smaller in the cells with inactive HIV. Since master regulators could be driving early transcriptional changes in the host T cells, enriched transcriptional factors (TFs) were searched for. In total, 58 TFs with enriched binding sites were found in promoters for DE genes expressed at 12 hpi. Of those, 10 with the highest enrichment values were selected for analysis. Some were up-regulated (*NFKB1* and *FOS*) and others down-regulated (*MYC*, *SIN3A*, *IRF1*, *ELK4*, *ETS1*, *YY1* and *E2F6*). As an exception, gene *EGR1* was the most up-regulated TF in HIV-infected cells but was found to be down-regulated in cells infected with unactivated HIV (Peng et al., 2014).

6.4.2 Hepatitis C Virus

Hepatitis C Virus (HCV) is the major causative agent for viral liver disease. However, the full impact of HCV infection has not been clearly established. The virus enters the liver cells through blood circulation; once the virus has infected cells, it causes tissue necrosis by a mechanism related to cytolysis, hepatic steatosis and oxidative stress. Protein variations by genotype change and influence those mechanisms, altering their effectiveness and success. Hence, the application of transcriptomic technologies such as RNA-seq is a powerful tool to characterize the transcriptional profile of the pathogen and serve as a guide for identification of its unique invasion mechanism (Irshad et al., 2013).

Characterization of human host cells' transcriptional response to hepatitis C infection has been done with RNA-seq applied with DE analysis in human hepatoma cells infected with HCV. The DE gene analysis identified 1,131 DE genes at 72 hpi and 700 at 96 hpi. From these, 172 genes had the most significant DE at both time points. Of these, 161 genes were up-regulated and 11 were down-regulated. Some up-regulated transcripts were identified as possible anti-HCV host response: *PIK3IP1*, *INPP5J*, *FAM46C* and *DDX60*. Additionally, the impact of HCV on biological pathways was measured. The MAPK signaling was the most altered pathway, as it had 10 overlapping genes, five of them forming part of a phosphatase subfamily (*DUSP1*, *DUSP4*, *DUSP8*, *DUSP10* and *DUSP16*). Their products act via dephosphorylation, negatively regulating MAPK proteins: *MAPK1*, *JNK*, *p38MAPK* and *ERK1*. The other four up-regulated genes during environmental stress are TF, such as *Gadd45A*, *Gadd45B*, *DDIT3* and *MRAS*. The adipocytokine signaling pathway is also overlapped by four up-regulated genes: *PPARGC1A* (which encodes a TF and acts on regulation of the energy metabolism); *IRS2* (encodes the insulin receptor substrate 2, a signaling molecule that acts as a molecular adaptation); *NFKBIA* (encodes a product protein capable of inhibiting NF-kappa-B/REL complexes involved with inflammation); and *ACSL* (which encodes a ligase that acts on lipid synthesis and fatty acid degradation). Another afflicted pathway was cell cycle, which was overlapped by up-regulated genes: *GADD45A* and *GADD45B* that encode TFs, *CDKN2B* that encodes an inhibitor of CDK kinases, and finally *CD14A* that encodes a dual-specificity protein proposed to regulate p53. Lastly, nitrogen metabolism was also enriched, with two overlapping genes related to amino acid biosynthesis: *CTH*, an encoding Cystathionine Gamma-Lyase, and *ASNS*, an encoding Glutamine-Dependent Asparagine Synthetase (Hojka-Osinska et al., 2016).

The immune system response activates the initial acute infection caused by the HCV, which is quite difficult to study, since most afflicted individuals are asymptomatic and do not seek medical care until they develop a serious condition. Even then, RNA-seq is such a powerful tool that it can characterize the early immune response corresponding to early acute infection of HCBV. Here, RNA-seq analysis was employed to characterize each transcriptomic interaction profile within three samples at distinct stages of infection. Using DE gene expression, 853 DE genes were detected at early acute infection. These small number of genes may be due to the small number of samples analyzed. These results lead to the conclusion that during acute infections, immune responses to HCV changes substantially the peripheral blood transcriptional signatures. An analysis of the differential regulation between blood transcriptome modules (BTMs) showed that 60 of these were differentially enriched and then classified by biological annotations. Gene expression directionality and dynamics were alike between BTMs of the same category. On the other hand, some BTMs were up-regulated as a response to early acute HCV infection, with related annotations like antiviral/interferon sensing, inflammatory/chemokines, monocytes, presentation of antigen and dendritic cell (DC) activation. While up-regulation of the BTM containing the T cell surface, activation (*M36*) annotation, showed a greater frequency in HCV-specific CD8[+] T cells. The response to acute HCV infection involves innate antiviral gene expression in peripheral immune cells. It is known that HCV triggers an IFN-mediated antiviral response in the liver cells, and BTMs associated with this response are strongly up-regulated upon acute HCV. To further characterize the BTM analysis, testing was done for enrichment of 277 gene-wide peripheral blood mononuclear cells interferon-stimulated gene set (PBMC ISG set). This set was significantly up-regulated on early and late acute infection, and the best defining ISGs were those implicated on the peripheral blood response against the viral infection (*CXCL10*), well-known ISGs (*Mx1, ISG15* and *OAS1*), ISGs related with immunomodulation (*CD38* and *IL15*) and ISG transcription factors (*IRF7, STAT1* and *STAT2*). This suggests that the innate immune response to HCV is not limited to the liver, but that it is also present in the peripheral blood, since the immune response in it has proven to be similar during infection (Rosenberg et al., 2018).

6.4.3 Coronaviruses

Coronaviruses are positive-sense, single-stranded RNA (ssRNA) viruses that infect mammals and birds. Two genera in particular, *Alphacoronavirus* and *Betacoronavirus*, cause respiratory diseases in humans. While the most common coronaviruses cause only mild respiratory diseases in humans, others induce more deadly infections (J. Cui et al., 2019). To date, three known highly virulent coronaviruses exist: SARS-CoV, MERS-CoV and the most recent SARS-CoV-2 (COVID-19), all of which cause respiratory syndromes and mortality to humans.

As their names indicate, both SARS-CoV viruses produce the disease known as acute respiratory distress syndrome (SARS). After acquisition, the virus infects the tracheobronchial and alveolar epithelial cells, causing severe lung injury via the own virus's cytopathic effect and the immune response to it (Gu & Korteweg, 2007). On the other hand, the Middle East respiratory syndrome virus (MERS) also infects the airway epithelial cells. The main characteristic of the MERS disease is a wider tissue tropism, the induction of a delayed inflammatory response and attenuation of innate immunity rendering it more lethal than the SARS coronaviruses (Choudhry et al., 2019).

Currently, a novel coronavirus named SARS-CoV-2, due to an 80% homology with the original SARS-CoV has spread to all over the world, provoking a pandemic with worse morbidity and mortality than the SARS epidemic in 2003. This novel coronavirus, most commonly referred to as COVID-19, originated from the Wuhan province in China in late 2019 and since has spread to 213 countries, producing more than one million confirmed infections worldwide. Similar to the other highly virulent coronaviruses, COVID-19 started as a zoonosis. Also like the other coronaviruses, it affects the host respiratory system, producing symptoms such as fever and dry cough, among others. Although it was considered to be less lethal compared to SARS and MERS, this virus is able to spread through person-to-person contact, and therefore it is easier to disperse inside whole communities than its previous counterparts (Petrosillo et al., 2020; Yuki et al., 2020).

Due to the current importance of the COVID-19 pandemic, a wide amount of both preliminary and full studies have been published in order to aid in the search for a treatment or to disclose its full nature in relation to other coronaviruses. This includes the use of transcriptomics to identify both pathogenicity factors and the main genes related to the host cell response to the infection in all its stages.

In order to achieve this, new studies have detailed the transcriptomic profiles of different cells affected by the infection. For instance, enrichment analysis of DE genes in both peripheral blood mononuclear cells (PBMC) and bronchoalveolar lavage fluid (BALF) during COVID-19 infection reveals several up-regulated genes related to immune responses and inflammation responses in PBMC cells and changes in membrane structures in the case of

BALF samples. In PBMC cells, 707 genes were found to be up-regulated and 316 down-regulated. In contrast, BALF samples showed 679 up-regulated and 325 down-regulated genes. Finally, BALF expression profiles revealed high expression of several cytokines related to the inflammatory response to COVID-19, such as *CCL2/MCP-1, CCL3/MIP-1A, CCL4/MIP1B, CXCL10/ IP-10* and *IL10*. Other cytokines (*CCL5, CXCL1, TNFSF10, TNFSF15, IL10, IL33, IL36G* and *IL36RN*) presented variable expression between patients. Finally, genes related to apoptosis routes, specifically the P53 pathway, were found to be altered within patients that showed reduced immune cell counts. The genes with altered expression included: *BIRC5, CCNB1, CCNB2, CDK1, CTSB, CTSD, CTSL, CTSZ, DDIT4, GTSE1, IGFBP3, NTRK1, RRAS, RRM2, STEAP3, TNFSF10* and *TP5313*. These findings indicated that COVID-19 infection in fact induces the apoptosis of immune cells (Xiong et al., 2020).

Expression analysis on samples from infected patients containing goblet and epithelial cells identified other up-regulated genes related to the host response to COVID-19. Some of the most notable up-regulated genes included the *ACE2* gene, which is the cellular receptor used by the virus to invade the host cells; the *IFI6* and *IFI27* interferon inducible proteins; the HERC6 Ubiquitin ligase that aids both the interferon-stimulated genes and the MHC-mediated antigen processing and cytokines *CCL8, CXCL10* and *CXCL11*. In this case, down-regulated genes involved cell development and growth-related genes, such as the *ALAS2*, an ALA-synthase found during development of red blood cells, and the *TMPRSS-11B* serine protease, which regulates the growth of lung cells (Butler et al., 2020). Similarly, interactome analysis based on transcriptomic profiles of infected 2B4 bronchial epithelial cells revealed more genes related to the host cells' responses against the virus. Such is the case of the *MCL1* gene, a positive regulator of apoptosis that is up-regulated in infected cells, resulting in either induction of the host cell or the immune cells in order to eliminate the infected cells at the early stages of infection. Another cell response to interrupt the virus infection is the down-regulation of *EEF1A1* elongation factor, involved in tRNA delivery during translation. In case of genes whose expression is directly affected by the virus, the *NUDFA10* protein gets down-regulated during infection, as a way of disrupting the host cell mitochondria. Another down-regulated gene is the *RNF128* ubiquitin ligase, which is related to antiviral response. Finally, one of the most important down-regulated genes by the host cells during coronavirus infection (including SARS and MERS) is the serine protease *TMPRSS2*, which is a protein that interacts directly with the viruses spike protein S, which is necessary for the virus to enter the cells (Guzzi et al., 2020). However, proof also exists that *TMPRSS2* expression is induced in the early stages of the infection, as well as down-regulation of *ACE2* (Sun et al., 2020).

The transcriptomic profile of COVID-19 has also been compared to the profiles of SARS, MERS, Ebola and H1N1 viruses. Considering only DE genes during cell infection, 358 DE genes significantly associated with COVID-19 were found within these transcriptomic profiles. Of these, three genes were considered highly significant (*SAA2*, *CCL20* and *IL8*). Furthermore, a gene enrichment analysis of DE genes associated with the host response brings up GO terms such as humoral immunity, leukocyte activation, neutrophil activation, tuberculosis response and cell death. The most significant pathways associated with COVID-19 infection highlighted by this analysis were cytokine response, *IL-17* signaling pathway, NF-Kb signaling and TNF signaling. Of the 358 associated DE genes mentioned before, 173 seem to be unique to COVID-19, SAA2 being the most significant. A GO analysis of these DE genes unique to COVID-19 demonstrated the link between certain genes (*CSF2*, *CSF3*, *ILIB* and *PTGS2*) to the *IL-17* signaling pathway, which induces the expression of antimicrobial peptides, cytokines and chemokines (Alsamman & Zayed, 2020). Other analyses have provided a major quantity of DE genes (2,002 DEGs specifically related to COVID-19 infection) in the transcriptomic profiles of patients. Of these, 771 were overexpressed and 1,231 were underexpressed. A GO term analysis for these genes revealed some of the most significantly related pathways to the overexpressed genes, such as innate immune response, immune response to viral infection, neutrophil activation, type-I interferon signaling and cytokine production. On the other hand, the underexpressed genes were mainly related to lymphocyte differentiation and T cell activation, indicating a suppression of T cells and activation of neutrophils as a response to COVID-19 infection. Comparing the COVID-19 transcriptomic profile to that from other viral infections, a couple of genes were found to be oppositely regulated: *ACO1*, a gene related to iron metabolism, is overexpressed in COVID-19 infections and underexpressed in other viral infections. Likewise, *ATL3*, a GTPase 3 required for the endoplasmatic reticulum membrane junctions is overexpressed in other viral infections, while it is underexpressed in COVID-19 infections, suggesting a role in the formation of viral replication sites (Thair et al., 2021).

Transcriptome profile analysis can also be used to identify viral factors that can be potentially targeted by drugs to secure an effective treatment. Research aimed toward this goal made a differential expression analysis in host cells targeted by COVID-19, such as human bronchial epithelial cells (NHBE) and adenocarcinomic human alveolar basal epithelial cells (A549). The analysis identified 143 DE genes for NHBE cells and 260 DE genes for A549 cells, having only 31 genes up-regulated in both cell types. Of these 31 genes, 19 were found to be common between different respiratory infection viruses, including SARS-CoV, MERS-CoV and Respiratory syncytial virus (RSV): *BCL2A1*, *CXCL2*, *CXCL3*, *EDN1*, *IFI6*, *IFI27*, *IFI44*, *IFIT1*, *IFITM1*, *IFIH1*, *IRF7*, *IRF9*, *MX1*, *OAS1*, *OAS2*, *PLSCR1*, *PTGS2* and *XAF1*. Most of these are interferon inducible proteins or regulatory factors, except for eight up-regulated genes found in both cell types, which are considered to be pro-viral host factors, since they are required for viral replication: *CFB*, *C1S*, *CXCL2*, *CXCL3*, *IFI44*,

TYMP, *PTGS2* and *XAF1*. Another couple of genes are also considered pro-viral host factors but are up-regulated only in NHBE cells: *NFKB1* and *TLR2* genes (Loganathan et al., 2020).

6.5 Human–Protozoan Interactions

A diversity of protozoan causes gastrointestinal diseases, skin, and nervous system infections such as toxoplasmosis, meningoencephalitis, and blood cell infections such as malaria (Todi & Srivastav, 2012). Multiple studies have been done with the objective of unveiling virulence factors or testing drugs for treatment. On the transcriptomic side, though, not many studies have been done, and some research has focused on trying to identify host–pathogen interactions. Some transcriptomic studies have been done on amoebae, but they were focused mainly on virulence factors and pathogenesis. Hence, this section focuses on the host–pathogen interactions in *Plasmodium falciparum* (malaria) and *Toxoplasma gondii* (toxoplasmosis), showing their most relevant gene expressions during infection (Figure 6.4).

6.5.1 Malaria

Approximately, 2.5 billion people are susceptible to become infected with malaria, which causes symptoms in 215 million people around the world. The infection spreads through mosquito bites, and after invading the organism, it

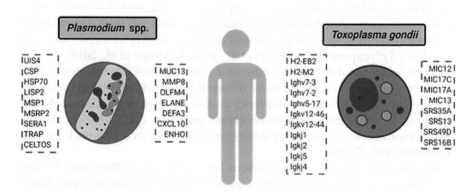

FIGURE 6.4
Main gene expression in host cells (blue) and protozoa (red) during infection. (Created with BioRender.com.)

causes a liver rupture due to the schizont stage of the parasite; the protozoan is then released as merozoites, which travel through the bloodstream into the heart and lung capillaries. Two types of malaria severity have been reported: (1) the complicated malaria shows signs of several organ tissue damage; and (2) uncomplicated malaria, which shows few or no signs of severe organ dysfunction (Mawson, 2013; Milner, 2018).

Scarce transcriptomics research has been done on host-protozoan interactions, since it is quite difficult to study the pathogen on its exoerythrocytic stage. Sporozoites produced by *Plasmodium* spp. form some kind of parasitophorous vacuole inside of infected hepatocytes, generating a direct relationship between the host cells and the parasite. This is why the application of transcriptomics would be very helpful to unveil such molecular interactions between the host immune cells and the parasite. With the application of dual RNA-seq, several genes related to the host defense mechanism against this infection have been identified. Infecting Huh7.5.1 hepatocytes with *Plasmodium berghei* has revealed that at the sporozoite stage, the most expressed transcripts are *UIS4* and *CSP*. On the other hand, *HSP70* and *LISP2* were the most expressed genes at 48 hpi. Both *UIS4* and *CSP* genes, along with *CELTOS*, were down-regulated at 48 hpi. At this same time point, merozoite genes such as *MSP1*, *MSRP2* and *SERA1* were up-regulated. These results agreed with those of microarray studies on *P. falciparum*, which showed some of the most highly produced transcripts in that organism at early infection, including five of the most expressed genes that were also at the top in the dual RNA-seq study (*CELTOS*, *TRAP*, *UI34*, *HSP70* and *CSP*) (LaMonte et al., 2019; Le Roch et al., 2003).

In the case of host factors possibly involved in the immune response against *P. falciparum*, some were identified as genes related to transcriptional pathways, regulators of energy homeostasis and others. Nonetheless, the most highly up-regulated transcript was the human mucosal innate immunity gene known as *MUC13*, which is also up-regulated against *Helicobacter pylori* infection. It seems that this overexpression is caused by *P. falciparum* to boost its evasion ability of the host immune system by covering its membrane with the *MUC13* glycoprotein (L. Cheng et al., 2016; LaMonte et al., 2019).

Genes down-regulated at the same time point (48 hpi) were related to host ribosome functions and DNA, while the most negatively regulated transcript was *ENHO*, which mainly intervenes in the positive regulation of Notch signaling. Another highly down-regulated gene was *CXCL10*, a chemokine ligand that acts in the α and γ interferon pathways as a hallmark protein (LaMonte et al., 2019).

Since malaria infections have various levels of severity, there are also different groups of DE genes between clinical groups of distinct severity. These DE gene groups have been identified through RNA-seq, with a total of 770 significantly DE genes between uncomplicated and severe malaria. The most notable genes that were highly expressed in severe malaria in comparison to mild malaria were: matrix metallopeptidase 8 (*MMP8*), olfactomedin 4 (*OLFM4*), neutrophil elastase (*ELANE*), defensing A3 (*DEFA3*) and all encoding granule proteins from neutrophils. GO analysis identified enriched specific biological functions in DE genes, such as immune response functions, cell motility and co-translational protein targeting (Lee et al., 2018).

6.5.2 Toxoplasmosis

A multitude of limitations exists in the pathogenesis of certain organisms. Some pathogens cannot disseminate from an initial point of infection and are limited to a specific tissue, and others can bypass certain biological barriers and travel through the circulatory system. The success of these infections depends on their colonization capacity. The main causative agent of toxoplasmosis, *Toxoplasma gondii*, roughly affects a third of the world population, causing a major chronic infection. Its ability to invade different tissues is one of the main factors that come into play when it comes to the infliction of a disease (Randall & Hunter, 2011). Perhaps, the application of RNA-seq and transcriptomics, in general, can shed some light on its capacity of tissue invasion and immune evasion.

Research has been done to study the transcriptional profiles of *T. gondii* and the host cells during the establishment of both acute and chronic infections. Using GO enrichment analysis, it was found that of the 100 most highly expressed genes from both types of infections, only 42 were shared. In GO terms, these genes were identified as related to translation, cellular metabolism, transcription and macromolecule biosynthesis. On the other hand, DE analysis established 547 DE genes between acute and chronic infections. From this total, 63 genes were > fivefold more abundant in acute infections in comparison to chronic ones. From these, 26 are hypothetical proteins, while the rest range from SAG-related sequence (SRS) anchored surface antigens (*SRS2/SRS29C*, *SRS20A*, *SAG1/SRS29B*, *SAG2/SRS34A*, *SRS52A* and *SRS54*) to rhoptry proteins (*ROP9*, *ROP16*, *ROP39* and *ROP40*). Further, 107 DE genes were associated with chronic infection. Unfortunately, 51 of those proteins are hypothetical. The others were annotated genes such as micronemes (*MIC12*, *MIC17C*, *MIC17A* and *MIC13*), glycolysis-related proteins (glucosephospate mutase, lactate dehydrogenase 2, glucose-6-phosphate isomerase and pyruvate kinase) and SRS genes (*SRS35A*, *SRS13*, *SAG2C/SRS49D* and *SRS9/SRS16B*). The DE analysis was also applied on acute versus uninfected samples, chronic versus uninfected and acute versus chronic at different time points. A total of 1,004 DE genes were expressed on acute versus uninfected, 2,510 in chronic versus uninfected and 1,872 DE genes were higher in chronically infected mice versus acutely infected. Interestingly, from the 1,004 genes abundant in acute versus uninfected, 902 were shared with the pathogen in the chronic versus uninfected. Also, 1,608 genes were up-regulated in the chronic versus uninfected time point that were not considered as DE in acute versus uninfected. This suggests that up-regulated genes that peak during acute infection are maintained during chronic infection when new genes start to be up-regulated. Applying GO enrichment analysis, it was seen that in the host cells, the overrepresented terms were those related to immune response and stress. When comparing DE genes between groups, notable differences were observed between acute and chronic infections, since several of the acute-infection specific genes included guanylate-binding proteins and GTPases induced by the interferon-γ pathway. In the case of abundant host genes during chronic infection, most of the genes were for immunoglobulin heavy chain regions, which suggest the use of different subsets of antibodies. Most of the DE genes in this group include *Ighv7-3, Ighv7-2, Ighv5-17, Igkv12-46, Igkv12-44, Igkj1, Igkj2, Igkj4* and *Igkj5*. Another highly expressed group of genes are *H2-EB2* and *H2-M2*, related to antigen presentation (Pittman et al., 2014).

6.6 Future Perspectives and Conclusions

Still, much research has to be done in the field of host–pathogen interactions. With technologies such as scRNA-seq and dual RNA-seq, it is now possible to characterize almost complete interactomes and the relationship between host–pathogen transcriptional profiles during infection. Furthermore, since it has been proven that infections caused by the same pathogen differ drastically in virulence or severity as a disease; these variations can be caused by both the transcriptional phenotype of pathogen strain or the particular host immune response to it. In order to truly identify such causes, it is essential to continue research on host–pathogen interactions with both traditional and dual RNA-seq. Knowledge on known interactomes could be misinterpreted or outdated, which makes the application of these technologies a mandatory practice to establish correct and fully fleshed transcriptional profiles between both interacting parts in these infectious diseases.

References

Alsamman, A. M., & Zayed, H. (2020). The transcriptomic profiling of COVID-19 compared to SARS, MERS, Ebola, and H1N1. *BioRxiv*, 2020.05.06.080960. doi:10.1101/2020.05.06.080960

Alspaugh, J. A. (2015). Virulence mechanisms and *Cryptococcus neoformans* pathogenesis. *Fungal Genetics and Biology, 78*(919), 55–58. doi:10.1016/j.fgb.2014.09.004

Aprianto, R., Slager, J., Holsappel, S., & Veening, J. W. (2016). Time-resolved dual RNA-seq reveals extensive rewiring of lung epithelial and pneumococcal transcriptomes during early infection. *Genome Biology, 17*(1). doi:10.1186/s13059-016-1054-5

Avital, G., Avraham, R., Fan, A., Hashimshony, T., Hung, D. T., & Yanai, I. (2017). scDual-Seq: Mapping the gene regulatory program of Salmonella infection by host and pathogen single-cell RNA-sequencing. *Genome Biology, 18*(1), 1–8. doi:10.1186/s13059-017-1340-x

Avraham, R., Haseley, N., Brown, D., Penaranda, C., Jijon, H. B., Trombetta, J. J., Satija, R., Shalek, A. K., Xavier, R. J., Regev, A., & Hung, D. T. (2015). Pathogen cell-to-cell variability drives heterogeneity in host immune responses. *Cell, 162*(6), 1309–1321. doi:10.1016/j.cell.2015.08.027

Baron, S. (1996). *Medical Microbiology.* (4th edition). University of Texas Medical Branch at Galveston.

Barquist, L., Westermann, A. J., & Vogel, J. (2016). Molecular phenotyping of infection associated small non-coding RNAs. *Philosophical Transactions of the Royal Society B: Biological Sciences, 371*(1707). doi:10.1098/rstb.2016.0081

Butler, D. J., Mozsary, C., Meydan, C., Danko, D., Foox, J., Rosiene, J., Shaiber, A., Afshinnekoo, E., MacKay, M., Sedlazeck, F. J., Ivanov, N. A., Sierra, M., Pohle, D., Zietz, M., Gisladottir, U., Ramlall, V., Westover, C. D., Ryon, K., Young, B., & Mason, C. E. (2020). Shotgun transcriptome and isothermal profiling of SARS-CoV-2 infection reveals unique host responses, viral diversification, and drug interactions. *BioRxiv,* 2020.04.20.048066. doi:10.1101/2020.04.20.048066

Chai, Q., Zhang, Y., & Liu, C. H. (2018). Mycobacterium tuberculosis: An adaptable pathogen associated with multiple human diseases. *Frontiers in Cellular and Infection Microbiology, 8*(MAY), 1–15. doi:10.3389/fcimb.2018.00158

Chang, S. T., Sova, P., Peng, X., Weiss, J., Law, G. L., Palermo, R. E., & Katze, M. G. (2011). Next-generation sequencing reveals HIV-1-mediated suppression of T cell activation and RNA processing and regulation of noncoding RNA expression in a CD4 + T cell line. *MBio, 2*(5), 1–9. doi:10.1128/mBio.00134-11

Chen, F., Zhang, C., Jia, X., Wang, S., Wang, J., Chen, Y., Zhao, J., Tian, S., Han, X., & Han, L. (2015). Transcriptome profiles of human lung epithelial cells A549 interacting with *Aspergillus fumigatus* by RNA-Seq. *PLoS ONE, 10*(8), 1–16. doi:10.1371/journal.pone.0135720

Chen, Y., Toffaletti, D. L., Tenor, J. L., Litvintseva, A. P., Fang, C., Mitchell, T. G., McDonald, T. R., Nielsen, K., Boulware, D. R., Bicanic, T., & Perfect, J. R. (2014). The *Cryptococcus neoformans* transcriptome at the site of human meningitis. *MBio, 5*(1), e01087–13. doi:10.1128/mBio.01087-13

Cheng, L., Mirko, R., Sara, L., Medea, P., Caroline, B., Eva, B., Myrthe, J., Bram, F., Wim, V. den B., Richard, D., Freddy, H., & Annemieke, S. (2016). The *Helicobacter heilmannii* hofE and hofF genes are essential for colonization of the gastric mucosa and play a role in IL-1β-induced gastric MUC13 expression. *Helicobacter, 21*(6), 504–522. doi:10.1111/hel.12307

Choudhry, H., Bakhrebah, M. A., Abdulaal, W. H., Zamzami, M. A., Baothman, O. A., Hassan, M. A., Zeyadi, M., Helmi, N., Alzahrani, F., Ali, A., Zakaria, M. K., Kamal, M. A., Warsi, M. K., Ahmed, F., Rasool, M., & Jamal, M. S. (2019). Middle East respiratory syndrome: Pathogenesis and therapeutic developments. *Future Virology, 14*(4), 237–246. doi:10.2217/fvl-2018-0201

Crossley, R. A., Gaskin, D. J. H., Holmes, K., Mulholland, F., Wells, J. M., Kelly, D. J., Van Vliet, A. H. M., & Walton, N. J. (2007). Riboflavin biosynthesis is associated with assimilatory ferric reduction and iron acquisition by *Campylobacter jejuni*. *Applied and Environmental Microbiology*, 73(24), 7819–7825. doi:10.1128/AEM.01919-07

Cui, J., Li, F., & Shi, Z. L. (2019). Origin and evolution of pathogenic coronaviruses. *Nature Reviews Microbiology*, 17(3), 181–192. doi:10.1038/s41579-018-0118-9

Dagenais, T. R. T., & Keller, N. P. (2009). Pathogenesis of *Aspergillus fumigatus* in invasive aspergillosis. *Clinical Microbiology Reviews*, 22(3), 447–465. doi:10.1128/CMR.00055-08

Eng, S. K., Pusparajah, P., Ab Mutalib, N. S., Ser, H. L., Chan, K. G., & Lee, L. H. (2015). Salmonella: A review on pathogenesis, epidemiology and antibiotic resistance. *Frontiers in Life Science*, 8(3), 284–293. doi:10.1080/21553769.2015.1051243

Griesenauer, B., Tran, T. M., Fortney, K. R., Janowicz, D. M., Johnson, P., Gao, H., Barnes, S., Wilson, L. S., Liu, Y., & Spinola, S. M. (2019). Determination of an interaction network between an extracellular bacterial pathogen and the human host. *MBio*, 10(3), 1–15. doi:10.1128/mBio.01193-19

Gu, J., & Korteweg, C. (2007). Pathology and pathogenesis of severe acute respiratory syndrome. *American Journal of Pathology*, 170(4), 1136–1147. doi:10.2353/ajpath.2007.061088

Guzzi, P. H., Mercatelli, D., Ceraolo, C., & Giorgi, F. M. (2020). Master regulator analysis of the SARS-CoV-2/human interactome. *Journal of Clinical Medicine*, 9(4), 982. doi:10.3390/jcm9040982

Hammerschmidt, S., Bergmann, S., Paterson, G. K., & Mitchell, T. J. (2007). Pathogenesis of *Streptococcus pneumoniae* infections: Adaptive immunity, innate immunity, cell biology, virulence factors. *Community-Acquired Pneumonia* (Issue January). doi:10.1007/978-3-7643-7563-8_8

Hojka-Osinska, A., Budzko, L., Zmienko, A., Rybarczyk, A., Maillard, P., Budkowska, A., Figlerowicz, M., & Jackowiak, P. (2016). RNA-seq-based analysis of differential gene expression associated with hepatitis C virus infection in a cell culture. *Acta Biochimica Polonica*, 63(4), 789–798. https://doi.org/10.18388/abp.2016_1343

Hovhannisyan, H., & Gabaldón, T. (2018). Transcriptome sequencing approaches to elucidate host–microbe interactions in opportunistic human fungal pathogens. *Current Topics in Microbiology and Immunology*. doi:10.1007/82

Humphrys, M. S., Creasy, T., Sun, Y., Shetty, A. C., Chibucos, M. C., Drabek, E. F., Fraser, C. M., Farooq, U., Sengamalay, N., Ott, S., Shou, H., Bavoil, P. M., Mahurkar, A., & Myers, G. S. A. (2013). Simultaneous transcriptional profiling of bacteria and their host cells. *PLoS ONE*, 8(12). doi:10.1371/journal.pone.0080597

Irshad, M., Mankotia, D. S., & Irshad, K. (2013). An insight into the diagnosis and pathogenesis of hepatitis C virus infection. *World Journal of Gastroenterology*, 19(44), 7896–7909. doi:10.3748/wjg.v19.i44.7896

Jia, X., Chen, F., Pan, W., Yu, R., Tian, S., Han, G., Fang, H., Wang, S., Zhao, J., Li, X., Zheng, D., Tao, S., Liao, W., Han, X., & Han, L. (2014). Gliotoxin promotes *Aspergillus fumigatus* internalization into type II human pneumocyte A549 cells by inducing host phospholipase D activation. *Microbes and Infection*, 16(6), 491–501. doi:10.1016/j.micinf.2014.03.001

Joshi, P. L., & Srivastava, R. K. (2009). *Training manual for medical officers* (D. G. of H. Services & M. of H. & F. Welfare, eds.). National Leprosy Eradication Programme. https://vimspawapuri.org/wp-content/uploads/2019/08/Leprosy_ Training_Manual.pdf

Kröger, C., Colgan, A., Srikumar, S., Händler, K., Sivasankaran, S. K., Hammarlöf, D. L., Canals, R., Grissom, J. E., Conway, T., Hokamp, K., & Hinton, J. C. D. (2013). An infection-relevant transcriptomic compendium for *Salmonella enterica serovar* Typhimurium. *Cell Host and Microbe, 14*(6), 683–695. doi:10.1016/j. chom.2013.11.010

LaMonte, G. M., Orjuela-Sanchez, P., Calla, J., Wang, L. T., Li, S., Swann, J., Cowell, A. N., Zou, B. Y., Abdel-Haleem Mohamed, A. M., Villa Galarce, Z. H., Moreno, M., Tong Rios, C., Vinetz, J. M., Lewis, N., & Winzeler, E. A. (2019). Dual RNA-seq identifies human mucosal immunity protein Mucin-13 as a hallmark of *Plasmodium exoerythrocytic* infection. *Nature Communications, 10*(1), 488. doi:10.1038/s41467-019-08349-0

Le Roch, K. G., Zhou, Y., Blair, P. L., Grainger, M., Moch, J. K., Haynes, J. D., De la Vega, P., Holder, A. A., Batalov, S., Carucci, D. J., & Winzeler, E. A. (2003). Discovery of gene function by expression profiling of the malaria parasite life cycle. *Science, 301*(5639), 1503–1508. doi:10.1126/science.1087025

Lee, H. J., Georgiadou, A., Walther, M., Nwakanma, D., Stewart, L. B., Levin, M., Otto, T. D., Conway, D. J., Coin, L. J., & Cunnington, A. J. (2018). Integrated pathogen load and dual transcriptome analysis of systemic host–pathogen interactions in severe malaria. *Science Translational Medicine, 10*(447), 1–31. doi:10.1126/ scitranslmed.aar3619

Lee, J., Lee, S. G., Kim, K. K., Lim, Y. J., Choi, J. A., Cho, S. N., Park, C., & Song, C. H. (2019). Characterisation of genes differentially expressed in macrophages by virulent and attenuated *Mycobacterium tuberculosis* through RNA-Seq analysis. *Scientific Reports, 9*(1), 1–9. doi:10.1038/s41598-019-40814-0

Li, H., Li, Y., Sun, T., Du, W., Li, C., Suo, C., Meng, Y., Liang, Q., Lan, T., Zhong, M., Yang, S., Niu, C., Li, D., & Ding, C. (2019). Unveil the transcriptional landscape at the Cryptococcus-host axis in mice and nonhuman primates. *PLoS Neglected Tropical Diseases, 13*(7), 1–24. doi:10.1371/journal.pntd.0007566

Loganathan, T., Ramachandran, S., Shankaran, P., Nagarajan, D., & Mohan S. S. (2020). Host transcriptome-guided drug repurposing for COVID-19 treatment: A meta-analysis based approach. *Peer J, 8*(April), e9357. doi:10.7717/peerj.9357

Maartens, G., Celum, C., & Lewin, S. R. (2014). HIV infection: Epidemiology, pathogenesis, treatment, and prevention. *Lancet, 384*(9939), 258–271. doi:10.1016/ S0140-6736(14)60164-1

Mawson, A. R. (2013). The pathogenesis of malaria: A new perspective. *Pathogens and Global Health, 107*(3), 122–129. doi:10.1179/2047773213Y.0000000084

Milner, D. A. (2018). Malaria pathogenesis. *Cole Spring Harbor Perspectives in Medicine, 8*(1), a025569.

Montoya, D. J., Andrade, P., Silva, B. J. A., Teles, R. M. B., Ma, F., Bryson, B., Sadanand, S., Noel, T., Lu, J., Sarno, E., Arnvig, K. B., Young, D., Lahiri, R., Williams, D. L., Fortune, S., Bloom, B. R., Pellegrini, M., & Modlin, R. L. (2019). Dual RNA-seq of human leprosy lesions identifies bacterial determinants linked to host immune response. *Cell Reports, 26*(13), 3574–3585.e3. doi:10.1016/j.celrep.2019.02.109

Muñoz, J. F., Delorey, T., Ford, C. B., Li, B. Y., Thompson, D. A., Rao, R. P., & Cuomo, C. A. (2018). Coordinated host–pathogen transcriptional dynamics revealed using sorted subpopulations and single, *Candida albicans* infected macrophages. *BioRxiv*, 350322. doi:10.1101/350322

Nuss, A. M., Beckstette, M., Pimenova, M., Schmühl, C., Opitz, W., Pisano, F., Heroven, A. K., & Dersch, P. (2017). Tissue dual RNA-seq allows fast discovery of infection-specific functions and riboregulators shaping host–pathogen transcriptomes. *Proceedings of the National Academy of Sciences of the United States of America*, 114(5), E791–E800. doi:10.1073/pnas.1613405114

Ortega, Á. D., Gonzalo-Asensio, J., & García-Del Portillo, F. (2012). Dynamics of Salmonella small RNA expression in non-growing bacteria located inside eukaryotic cells. *RNA Biology*, 9(4), 469–488. doi:10.4161/rna.19317

Peng, X., Sova, P., Green, R. R., Thomas, M. J., Korth, M. J., Proll, S., Xu, J., Cheng, Y., Yi, K., Chen, L., Peng, Z., Wang, J., Palermo, R. E., & Katze, M. G. (2014). Deep sequencing of HIV-infected cells: Insights into nascent transcription and host-directed therapy. *Journal of Virology*, 88(16), 8768–8782. doi:10.1128/jvi.00768-14

Petrosillo, N., Viceconte, G., Ergonul, O., Ippolito, G., & Petersen, E. (2020). COVID-19, SARS and MERS: Are they closely related? *Clinical Microbiology and Infection*, 26(6), 729–734. doi:10.1016/j.cmi.2020.03.026

Pisu, D., Huang, L., Grenier, J. K., & Russell, D. G. (2020). Dual RNA-seq of Mtb-infected macrophages in vivo reveals ontologically distinct host–pathogen interactions. *Cell Reports*, 30(2), 335–350.e4. doi:10.1016/j.celrep.2019.12.033

Pittman, K. J., Aliota, M. T., & Knoll, L. J. (2014). Dual transcriptional profiling of mice and *Toxoplasma gondii* during acute and chronic infection. *BMC Genomics*, 15(1), 1–19. doi:10.1186/1471-2164-15-806

Randall, L. M., & Hunter, C. A. (2011). Parasite dissemination and the pathogenesis of toxoplasmosis. *European Journal of Microbiology and Immunology*, 1(1), 3–9. doi:10.1556/eujmi.1.2011.1.3

Romo, J. A., Zhang, H., Cai, H., Kadosh, D., Koehler, J., Saville, S., & Lopez-Ribot, J. L. (2019). Global transcriptomic analysis of the *Candida albicans* response to treatment with a novel inhibitor of filamentation. *Msphere*, 4(5), e00620-19.

Ronald, A. R., & Plummer, F. A. (1985). Chancroid and *Haemophilus ducreyi*. *Annals of Internal Medicine*, 102(5), 706–708. doi:10.7326/0003-4819-102-5-705

Rosenberg, B. R., Depla, M., Freije, C. A., Gaucher, D., Mazouz, S., Boisvert, M., Bédard, N., Bruneau, J., Rice, C. M., & Shoukry, N. H. (2018). Longitudinal transcriptomic characterization of the immune response to acute hepatitis C virus infection in patients with spontaneous viral clearance. *PLoS Pathogens*, 14(9), 1–24. doi:10.1371/journal.ppat.1007290

Saliba, A. E., Santos, C. S., & Vogel, J. (2017). New RNA-seq approaches for the study of bacterial pathogens. *Current Opinion in Microbiology*. doi:10.1016/j.mib.2017.01.001

Sun, J., Ye, F., Wu, A., Yang, R., Pan, M., Sheng, J., Zhu, W., Mao, L., Wang, M., Huang, B., Tan, W., & Jiang, T. (2020). Comparative transcriptome analysis reveals the intensive early-stage responses of host cells to SARS-CoV-2 infection. *BioRxiv*, 2020.04.30.071274. doi:10.1101/2020.04.30.071274

Thair, S. A., He, Y. D., Hasin-Brumshtein, Y., Sakaram, S., Pandya, R., Toh, J., & Sweeney, T. E. (2021). Transcriptomic similarities and differences in host response between SARS-CoV-2 and other viral infections. *IScience*, 24(1), 101947.

Todi, S., & Srivastav, O. (2012). Infections in the immunocompromised host. In *ICU Protocols: A Stepwise Approach* (3rd edition, pp. 425–429). India: Springer. doi:10.1007/978-81-322-0535-7_54

Watkins, T. N., Liu, H., Chung, M., Hazen, T. H., Dunning Hotopp, J. C., Filler, S. G., & Bruno, V. M. (2018). Comparative transcriptomics of *Aspergillus fumigatus* strains upon exposure to human airway epithelial cells. *Microbial Genomics, 4*(2). doi:10.1099/mgen.0.000154

Xiong, Y., Liu, Y., Cao, L., Wang, D., Guo, M., Jiang, A., Guo, D., Hu, W., Yang, J., Tang, Z., Wu, H., Lin, Y., Zhang, M., Zhang, Q., Shi, M., Liu, Y., Zhou, Y., Lan, K., & Chen, Y. (2020). Transcriptomic characteristics of bronchoalveolar lavage fluid and peripheral blood mononuclear cells in COVID-19 patients. *Emerging Microbes and Infections, 9*(1), 761–770. doi:10.1080/22221751.2020.1747363

Yuki, K., Fujiogi, M., & Koutsogiannaki, S. (2020). COVID-19 pathophysiology: A review. *Lancet, 395*(April), 1315.

Zanto, T. P., Hennigan, K., Östberg, M., Clapp, W. C., & Gazzaley, A. (2010). Pathogenesis of genital tract disease due to *Chlamydia trachomatis*. *Journal of Infectious Diseases, 46*(4), 564–574. doi:10.1016/j.cortex.2009.08.003.Predictive

Index

Page numbers in *italics* denote figures; those in **bold** denote tables.